Professor Murray Gell-Mann is one of the most influential and brilliant scientists of the twentieth century. His work on symmetries, including the invention of the 'quark', in the 1950s and early 1960s has provided a foundation for much of modern particle physics and was recognised by the award of the Nobel Prize for Physics in 1969.

This book is a collection of research articles especially written by eminent scientists to celebrate Gell-Mann's 60th birthday, in September 1989. The main body of contributions are concerned with theoretical particle physics and its applications to cosmology.

Elementary Particles and the Universe

Elementary Particles and the Universe

ESSAYS IN HONOR OF *Murray Gell-Mann*

Edited by **John H. Schwarz**

Harold Brown Professor of Theoretical Physics,
California Institute of Technology

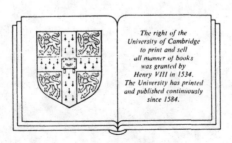

CAMBRIDGE UNIVERSITY PRESS
CAMBRIDGE
NEW YORK   PORT CHESTER   MELBOURNE   SYDNEY

CAMBRIDGE UNIVERSITY PRESS
Cambridge, New York, Melbourne, Madrid, Cape Town, Singapore, São Paulo

Cambridge University Press
The Edinburgh Building, Cambridge CB2 2RU, UK

Published in the United States of America by Cambridge University Press, New York

www.cambridge.org
Information on this title: www.cambridge.org/9780521412537

© Cambridge University Press 1991

This book is in copyright. Subject to statutory exception
and to the provisions of relevant collective licensing agreements,
no reproduction of any part may take place without
the written permission of Cambridge University Press.

First published 1991
This digitally printed first paperback version 2005

A catalogue record for this publication is available from the British Library

ISBN-13 978-0-521-41253-7 hardback
ISBN-10 0-521-41253-6 hardback

ISBN-13 978-0-521-01759-6 paperback
ISBN-10 0-521-01759-9 paperback

Contents

Preface
JOHN H. SCHWARZ — ix

Excess Baggage
JAMES B. HARTLE — 1

Through the Clouds
EDWARD WITTEN — 17

Covariant Formulations of the Superparticle and the Superstring
LARS BRINK — 23

Chiral Symmetry and Confinement
T. GOLDMAN — 41

The Original Fifth Interaction
YUVAL NE'EMAN — 47

The Mass Hierarchy of Leptons and Quarks as a New Symmetry
HARALD FRITZSCH — 61

Spacetime Duality in String Theory
JOHN H. SCHWARZ — 69

Supersymmetry and Quasi-Supersymmetry
Y. NAMBU — 89

The Exceptional Superspace and the Quadratic Jordan Formulation of Quantum Mechanics
M. GÜNAYDIN — 99

Algebra of Reparametrization-Invariant and Normal Ordered Operators in Open String Field Theory
P. RAMOND — 121

Superconductivity of an Ideal Charged Boson System
T. D. LEE — 135

Some Remarks on the Symmetry Approach to Nuclear Rotational Motion
L. C. BIEDENHARN AND P. TRUINI — 157

Uncomputability, Intractability and the Efficiency of Heat Engines
SETH LLOYD — 175

The New Mathematical Physics
I. M. SINGER — 185

"Is Quantum Mechanics for the Birds?"
V. L. TELEGDI   **193**

The Gell-Mann Age of Particle Physics
ABDUS SALAM   **207**

Remarks on the occasion of Murray Gell-Mann's more or less 60th Birthday
M. GOLDBERGER   **211**

Preface

A two-day symposium in celebration of Murray Gell-Mann's 60th birthday was held at the California Institute of Technology on January 27–28, 1989. The theme of the Symposium was "Where are Our Efforts Leading?" Each speaker was asked to choose one (or more) of the great challenges in science or human affairs and try to answer, in connection with our present effort to respond to that challenge, "Where do we stand? What kind of progress are we making? In fifty or a hundred years, how do you think today's efforts will appear?". The topics discussed spanned a very broad range, representative of Murray's remarkably diverse interests and activities. These included particle physics and quantum cosmology, studies of complex adaptive systems, environmental challenges and studies, education and equality of opportunity, arms control and governmental issues.

Given the unusually broad scope of the Symposium, we decided it would be appropriate to publish separately a 'physics volume,' including all of the more technical contributions in theoretical physics and related topics. There were many marvelous contributions in other areas that we hope to publish elsewhere. The present volume includes the texts of presentations at the Symposium by Professors J. B. Hartle, E. Witten, H. Fritzsch, T. D. Lee, I. M. Singer, and V. L. Telegdi. It also includes ten additional contributed papers by well-known physicists who are close personal friends of Murray Gell-Mann. In addition, some brief personal remarks have been extracted from speeches given at the Symposium by Professors A. Salam and M. L. Goldberger.

John H. Schwarz
November 1990

EXCESS BAGGAGE

James B. Hartle
Department of Physics
University of California
Santa Barbara, CA 93106

1 QUANTUM COSMOLOGY

It is an honor, of course, but also a pleasure for me to join in this celebration of Murray Gell-Mann's sixtieth birthday and to address such a distinguished audience. Murray was my teacher and more recently we have worked together in the search for a quantum framework within which to erect a fundamental description of the universe which would encompass all scales — from the microscopic scales of the elementary particle interactions to the most distant reaches of the realm of the galaxies — from the moment of the big bang to the most distant future that one can contemplate. Such a framework is needed if we accept, as we have every reason to, that at a basic level the laws of physics are quantum mechanical. Further, as I shall argue below, there are important features of our observations which require such a framework for their explanation. This application of quantum physics to the universe as a whole has come to be called the subject of quantum cosmology.

The assignment of the organizers was to speak on the topic "Where are our efforts leading?" I took this as an invitation to speculate, for I think that it is characteristic of the frontier areas of science that, while we may know what direction we are headed, we seldom know where we will wind up. Nevertheless, I shall not shrink from this task and endeavor, in the brief time available, to make a few remarks on the future quantum cosmology. The point of view that I shall describe owes a great deal to my conversations with Murray.

One cannot contemplate the history of physics without becoming aware that many of its intellectual advances have in common that some idea which was previously accepted as fundamental, general and inescapable was subsequently seen to be consequent, special and dispensible. Further, this was often for the following reason: The idea was not truly a general feature of the world, but only *perceived* to be general because of our special place in the universe and the limited range of our experience. In fact, it arose from a true physical fact but one which is a special situation in a yet more general theory. To quote Murray himself from a talk of several years ago: "In my field an important new idea ..... almost always includes a negative statement, that some previously accepted principle is unnecessary and can be dispensed with. Some earlier correct idea .... was accompanied by unnecessary intellectual baggage and it is now necessary to jettison that baggage."[1]

In cosmology it is not difficult to cite previous examples of such transitions.[2] The transition from Ptolemaic to Copernican astronomy is certainly one. The centrality of the earth was a basic assumption of Ptolemaic cosmology. After Copernicus, the earth was seen not to be fundamentally central but rather one planet among others in the solar system. The earth was in fact distinguished, not by a law of nature, but rather by our own position as observers of the heavens. The idea of the central earth was excess baggage.

The laws of geometry of physical space in accessible regions obey the Euclidean laws typified by the Pythagorean theorem on right triangles. Laws of physics prior to 1916, for example those governing the propagation of light, incorporated Euclidean geometry as a fundamental assumption. After Einstein's 1916 general theory of relativity, we see Euclidean geometry not as fundamental, but rather as one possibility among many others. In Einstein's theory, the geometry in the neighborhood of any body with mass is non-Euclidean and this curvature gives a profound geometrical explanation for the phenomena we call gravity. On the very largest scales of cosmology, geometry is significantly curved and it is the dynamics of this geometry which describes the evolution of the universe. Euclidean geometry is the norm for us, not because it is fundamental, but only because our observations are mostly local and because we happen to be living far from objects like black holes or epochs like the big bang. The idea of a fixed geometry was excess baggage.

In each of these examples there was a feature of the current theoretical framework which was perceived as fundamental but which in truth was a consequence of our particular position and our particular time in the universe in a more general theoretical framework. Our description was too special. There was excess baggage which had to be discarded to reach a more general and successful viewpoint. Thus, in our effort to predict the future of our quantum cosmology, the question naturally arises: Which features of our *current* theoretical framework reflect our special position in the universe and which are fundamental? Which are excess baggage?

We live at a special position in the universe, not so much in place, as in time. We are late, living some ten billion years after the big bang, a time when many interesting possiblities for physics could be realized which are not easily accessible now. Moreover, we live in a special universe. Ours is a universe which is fairly smooth and simple on the largest scales, and the evidence of the observations is that if we look earlier in time it is smoother and simpler yet. There are simple initial conditions which are only one very special possibility out of many we could imagine.

The question I posed above can therefore be generalized: Which features of our current theoretical framework are fundamental and which reflect our special positon in the universe or our special initial conditions. Are there natural candidates for elements of the framework which might be generalized? Is there excess baggage? This is the question that I would like to address today for quantum cosmology.

2 QUANTUM MECHANICS

First, let us review the current quantum mechanical framework today as it was developed in the twenties, thirties and forties, and as it appears in most textbooks today.

It is familiar enough that there is no certainty in this world and that therefore we must deal in probabilities. When probabilities are sufficiently good we act. It wasn't certain that my plane to Los Angeles wouldn't be blown up, but, because I thought that the probability was sufficiently low, I came. Experimentalists can't be certain that the results of their measurements are not in error, but, because they estimate the probability of this as low, they publish. And so on. It was the vision of classical physics that fundamentally the world *was* certain and that the use of probablility was due to lack of precision. If one looked carefully enough to start, one could with certainty predict the future. Since the discovery of quantum mechanics some sixty years ago, we have known that this vision is only an approximation and that probabilities are inevitable and fundamental. Indeed,

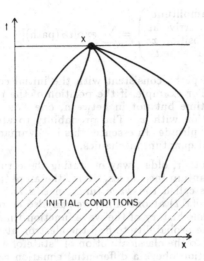

Fig. 1: In quantum mechanics the probability for a particle to be observed at position x at time t is the absolute square of the amplitude to arrive there. In Feynman's sum over histories formulation, that amplitude is the sum over all paths which are consistent with the initial conditions and arrive at x at t of the amplitude for each path $\exp[iS(\text{path})]$. In the Hamiltonian formulation of quantum mechanics, this sum is called the wave function.

the transition from classical to quantum physics contains a good example of excess baggage. In this case it was classical determinism. Classical evolution was but one possibility out of many although in the great majority of situations we were used to by far the most probable evolution.

Quantum mechanics constructs probabilities according to characteristic rules. As a simple example let us consider the the quantum mechanics of a slowly moving particle such as an electron in a solid. Its path in spacetime is called its history.

In classical mechanics the time history of the electron is a definite path determined by the electron's initial conditions and its equation of motion. Thus, given sufficiently precise initial conditions, a later observation of position can yield only a single, certain, predictable result. In quantum mechanics, given the most precise possible initial conditions, all paths are possible and all possible results of the observation may occur. There is only a probability for any one of them. The probability of an observation of position at time t yielding the value x, for example, is constructed as follows:

A complex number, called the amplitude, is assigned to each path. This number has the form $\exp[iS(\text{path})]$ where S — the action — summarizes the inertial properties of the electron and its interactions. The amplitude to arrive at x at t is the sum of the path amplitudes over all paths which are consistent with the initial conditions and which end at x at t (Figure 1).

$$\begin{pmatrix} \text{amplitude} \\ \text{to arrive at} \\ \text{position } x \\ \text{at time } t \end{pmatrix} = \sum \exp[iS(\text{path})] \quad . \tag{1}$$

The sum here is over all paths consistent with the initial conditions and ending at position x at time t. For example, if the position of the particle were actually measured at a previous time but not in between, one would sum over all paths which connect that position with x. The probability to arrive at x at t is the absolute square of this amplitude. In essence, this is Feynman's sum over histories formulation of the rules of quantum mechanics.

There is another, older, way of stating these rules called the Hamiltonian formulation of quantum mechanics. Here, the amplitude for observing the election at x at time t is called the wave function $\psi(x,t)$. If, as above, we can construct the amplitude $\psi(x,t)$ at one time, we can also employ this construction at all other times. The time history of the wave function time history gives a kind of running summary of the probability to find the electron at x. The wave function is thus the closest analog to the classical notion of "state of a system at a moment of time." The wave function obeys a differential equation called the Schrödinger equation

$$i\frac{\partial \psi}{\partial t} = H\psi \tag{2}$$

where H is an operator, called the Hamiltonian, whose form can be derived from the action. It summarizes the dynamics in an equivalent way. Thus, although the electron's position does not evolve according to a deterministic rule, its wave function does.

Feynman's sum over histories formulation of quantum mechanics is fully equivalent to the older Hamiltonian formulation for this case of slowly moving particles. One can calculate the wave function from (1) for two nearby times and demonstrate that it satisfies the Schrödinger equation (2). One can use either formulation of quantum mechanics in this and many other situations as well.

More generally than the probability of one observation of position at one time, the probabilities of time *sequences* of observations are of interest. Just such a sequence in needed, for example, to check whether or not the electron is moving along a classical path between the initial conditions and time t. The amplitude for positive answers to checks of whether the electron is located inside a sequence of position intervals is the sum of the amplitude for a path over all paths which thread these intervals (Figure 2). The joint probability for the sequence of positive answers is the square of this amplitude.

This means that the probability of finding the particle at x at t depends not only on the initial conditions, but also on what measurements were carried out betweem the initial conditions and t. Even if one doesn't know the results of these measurements, the probabilities at t will still depend on whether or not they were carried out. Thus, there are in quantum mechanics different rules for evolution depending on whether measurements occurred or did not. In the Hamiltonian formulation this means that the wave function evolves by the Schrödinger equation only in between measurements. At a measurement it evolves by a different rule – the notorious "reduction of the wave packet".

In some circumstances it makes no difference to the probabilities whether prior measurements were carried out or not. For example, the classical

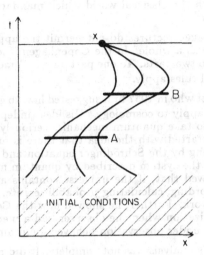

Fig. 2: Checks may be carried out as to whether the particle passed through position intervals A, B, $\cdots$ at times in between the initial conditions and t. The amplitude to arrive at position x at time t with a sequence of positive answers to these checks is the sum over all paths which meet the initial conditions and arrive at x and t having threaded the intervals A, B, $\cdots$ The probability to arrive at x at t will depend on whether these checks are carried out *even if nothing is known of the results*.

limit of quantum mechanics occurs when the initial conditions are such that only a single path to x – the classical one– contributes to the sum over histories. Then, as long as sufficiently crude measurements are considered, it makes no difference to the probability of x at t whether they are made or not. The electron follows the classical evolution. In such cases the classical history is said to "decohere".

3 FROM BOHR, TO EVERETT, TO POST-EVERETT

The framework of quantum mechanics described in the previous section was the starting point for the "Copenhagen" interpretations of this subject. An idea characteristic of the Copenhagen interpretations was that there was something external to the framework of wave function and Schrödinger equation which was necessary to interpret the theory. Various expositors put this in different ways: Bohr[3] spoke of alternative descriptions in terms of classical language. Landau and Lifshitz[4] emphasized preferred classical observables. Heisenberg and others[5] stressed the importance of an external observer for whom the wave function was the most complete summary possible of information about an observed system. All singled out the measurement process for a special role in the theory. In various ways, these authors were taking as fundamental the manifest existence of the classical world that we see all about us. That is, they were taking as fundamental the existence of objects which do have histories obeying classical probability rules and, except for the occasional intervention of the quantum world as in a measurement, obey deterministic classical equations of motion. This existence of a classical world, however it was phrased, was an important part of the Copenhagen interpretations for it was the contact with the classical world which mandated the "reduction of the wave packet".

The Copenhagen pictures do not permit the application of quantum mechanics to the universe as a whole. In the Copenhagen interpretations the universe is always divided into two parts: To one part quantum mechanical rules apply. To the other part classical rules apply.[6]

It was Everett who in 1957 first suggested how to generalize the Copenhagen framework so as to apply to cosmology.[7] His idea (independently worked out by Murray in 1963) was to take quantum mechanics seriously and apply it to the universe as a whole. He started with the idea that there is one wave function for the universe always evolving by the Schrödinger equation and never reduced. Any observer would be part of the system described by quantum mechanics, not seperate from it. Everett showed that, if the universe contains an observer, then its activities –measuring, recording, calculating probabilities, etc.– could be described in this generalized framework. Further, he showed how the Copenhagen structure followed from this generalization when the observer had a memory which behaved classically and the system under observation behaved quantum mechanically.

Yet, Everett's analysis was not complete. It did not explain the manifest existence of the classical world much less the existence of something as sophisticated as an observer with a classically behaving memory. Classical behavior, after all, is not generally exhibited by quantum systems. The subsequent, post-Everett, analysis of this question involves a synthesis of the ideas of many people. Out of many, I might mention in particular the work of Joos and Zeh,[8] Zurek,[9] Griffiths,[10] and latterly Murray, Val Telegdi, and myself. It would take us too far to attempt to review the mechanisms by which the classical world arises but we can identify the theoretical feature to which its origin can, for the most part, be traced.

A classical world cannot be a general feature of the quantum mechanics of the universe, for the number of states which imply classical behavior in any sense is but a poor fraction of the total states available to the universe. Classical behavior, of course, can be an approximate property of a *particular* state as for a particle in a wave packet whose center moves according to the classical equations of motion. But, the universe *is* in a particular state (or a particular statistical mixture of states). More exactly, particular quantum initial conditions must be posed to make any prediction in quantum cosmology. It is to the particular features of the initial conditions of the universe, therefore, that we trace the origin of today's classical world and the possibility of such information gathering and utilizing systems as ourselves.

In retrospect, the Copenhagen idea that a classical world or an observer together with the act of measurement occupy a fundamentally distinguished place in quantum theory can be seen to be excess baggage. These ideas arose naturally, in part from our position in the late universe where there *are* classically behaving objects and even observers. In part, they arose from the necessary focus on laboratory experiments as the most direct probe of quantum phenomena where there *is* a clear distinction between observer and observed. However, these features of the world, while true physical facts in these situations, are not fundamental. Quantum mechanically, the classical world and observers are but some possible systems out of many and measurements are but one possible interaction of such systems out of many. Both are unlikely to exist at all in the very early universe. They are, in a more general post-Everett framework, *approximate* features of the late universe arising from its particular quantum state. The classical reality to which we have become so attached by evolution is but an approximation in an entirely quantum mechanical world made possible by specific initial conditions.

The originators of the Copenhagen interpretation were correct; something beyond the wave function and the Schrödinger equation *is* needed to interpret quantum mechanics. But, that addition is not an external restriction of the domain to which the theory applies. It is the initial conditions of the universe specified within the quantum theory itself.

4 ARROWS OF TIME

Heat flows from hot to cold bodies, never from cold to hot. This is the essence of the second law of thermodynamics that entropy always increases. In the nineteeth century this law was thought to be strict and fundamental.[11] A direction of time was distinguished by this law of nature. After the success of the molecular theory of matter, with its *time symmetric* laws of dynamics, the increase in entropy was seen to be *approximate* – one possibility out of many others although the overwhelmingly most probable possibility in most situations. A direction of time was distinguished, not fundamentally, but rather by our position in time relative to initial conditions of simplicity and our inability to follow molecular motion in all accuracy. The idea of a fundamental thermodynamic arrow of time was excess baggage.

Exchanging the thermodynamic arrow of time for simple initial conditions might seem to be exchanging one asymmetry for another. "Why", one could ask, "was the universe simple in the past and not in the future?" In fact, this is not a question. There is no way of specifying that direction in time which is the past except by calling it the direction in which the universe is simple. Its not an arrow of time which is fundamental, but rather the fact that the universe is simple at one end and not at the other.[12]

The Hamiltonian formulation of quantum mechanics also posseses a similar distinction between past and future. From the knowledge of the state of a system at one time *alone*, one can predict the future but one cannot, in general, retrodict the past.[13] This is an expression of causality in quantum theory. Indeed, quantum mechanics in general prohibits the construction of history. The two slit experiment is a famous example (Figure 3). Unless there was a measurement, it is not just that one can't say for certain which slit the electron went through; it is meaningless even to assign a probability. In this asymmetry between past and future the notion of "state" in quantum mechanics is different from that of classical physics from which both future and past can be extrapolated. In Hamiltonian quantum mechanics, the past is the direction which is comprehensible by theory; the future is the direction which is generally predictable.

As long as the distinction between future and past in cosmology is fundamental, it is perhaps reasonable to formulate a quantum cosmology using a quantum mechanics which maintains this same distinction as fundamental. However, it would seem more natural if causality were an empirical conclusion about the universe rather than a prerequisite for formulating a theory of it. That is, it would seem more natural if causality were but one of many options available in quantum mechanics, but the one appropiate for this particular universe. For example, one might imagine a universe where both initial *and* final conditions are set. The generalization of Hamiltonian quantum mechanics necessary to accommodate such situations is not difficult to find.[14] In fact, it is ready to hand in the sum over histories formulation described briefly in Section II. The rules are the same, but both initial and final conditions are imposed on a contributing history. In the context of such a generalization, the arrow of time in quantum mechanics, and the associated

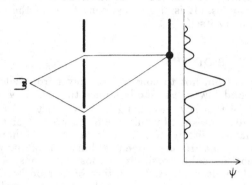

Fig. 3: The two slit experiment. An electron gun at right emits an electron traveling towards a screen with two slits, its progress in space recapitulating its evolution in time. From a knowledge of the wave function at the detecting screen alone, it is not possible to assign a probability to whether the electron went through the upper slit or the lower slit.

notion of state,[15] would not be fundamental. Rather they would be features of the theory arising from the fact that those conditions which fix our particular universe are comprehensible at one of its ends but not at the other.

5 QUANTUM SPACETIME

Gravity governs the structure and evolution of the universe on the largest scales, both in space and time. To pose a quantum theory of cosmology, therefore, we need a quantum theory of gravity. The essence of the classical theory of gravity – Einstein's general relativity – is that gravity is curved spacetime. We thus need a quantum theory of space and time themselves in which the geometry of spacetime will exhibit quantum fluctuations.

Finding a consistent, managable quantum theory of gravity has been one of the goals of theoretical research for the last thirty years. The chief problem has been seen as finding a theory which puts out more predictions then there are input parameters. But, in cosmology there is also a conflict with the existing framework of quantum mechanics. These difficulties are called "the problem of time".[16]

Time plays a special and peculiar role in the familiar Hamiltonian formulation of quantum mechanics. All observations are assumed to be unambiguously characterized by a single moment of time and we calculate probabilities for "observations at one moment of time". Time is the only observable for which there are no interfering alternatives (as a measurement of momentum is an interfering alternative for a measurement of position). Time is the sole observable not represented in the theory as an operator but rather enters the Schrödinger equation as a parameter describing evolution.

If the geometry of spacetime is fixed, external and classical, it provides the preferred time of quantum mechanics. (More exactly, a fixed background spacetime provides a family of timelike directions equivalent in their quantum mechanics because of relativistic causality.) If the geometry of spacetime is quantum mechanical – fluctuating and without definite value – then it cannot supply a notion of time for quantum mechanics. There is thus a conflict between the familiar Hamiltonian quantum mechanics with a preferred time and a quantum theory of spacetime. This is the "problem of time".

The usual response to this difficulty has been to keep Hamiltonian quantum mechanics but to give up on spacetime. There are several candidates for a replacement theory. Perhaps space and time are seperate quantum mechanically with the beautiful synthesis of Minkowski and Einstein emerging only in the classical limit.[18] Perhaps spacetime needs to be augmented by other, now hidden, variables which play the role of a preferred time in quantum mechanics.[19] Perhaps spacetime is simply a totally inappropriate notion fundamentally and emerges only in the classical limit of some yet more subtle theory.

Each of these ideas could be right. I would like to suggest, however, that there is another alternative. This is that the notion of a preferred time in quantum mechanics is another case of excess baggage. That the Hamiltonian formulation of quantum mechanics is simply not general enough to encompass a quantum theory of spacetime. That the present formulation of quantum mechanics is but an approximation to a more general framework appropiate because of our special position in the universe.

It is not difficult to identify that property of the late universe which would have led to the perception that there is a preferred time in quantum mechanics. Here, now, on all accessible scales, from the smallest ones of the most energetic accelerators to the largest ones of the furthest seeing telescopes, spacetime *is* classical. The preferred time of quantum mechanics reflects this true physical fact. However, it can only be an approximate fact if spacetime is quantum mechanical. Indeed, like all other aspects of the classical world, it must be a property of the particular quantum initial conditions of the universe. The present formulation of quantum mechanics thus may be only an approximation to a more general framework appropiate because of particular initial conditions which mandate classical spacetime in the late universe.

What is this more general quantum mechanics? Feynman's sum over histories supplies a natural candidate.[20,21,22] I shall now sketch why. We saw earlier that Feynman's approach was equivalent to the Hamiltonian one for particle quantum mechanics. It is also for field theory. A little thought about our example shows, however, that this equivalence arises from a special property of the histories – that they do not double back on themselves in time. Thus, the preferred time enters the sum over histories formulation as a restriction on the allowed histories.

In a quantum theory of spacetime the histories are four dimensional spacetimes geometries $\mathcal{G}$ with matter fields $\phi(x)$ living upon them. The sums over histories defining quantum amplitudes therefore have the form:

$$\sum_{\mathcal{G}} \sum_{\phi(x)} \exp\bigl(iS[\mathcal{G}, \phi(x)]\bigr) \qquad (3)$$

where S is the action for gravity coupled to matter. The sums are restricted by the theory of inial conditions and by the observations whose probabilities one aims to compute.

The interior sum in (3)

$$\sum_{\phi(x)} \exp\left(iS[\mathcal{G}, \phi(x)]\right) \tag{4}$$

defines a usual quantum field theory in a temporarily fixed background spacetime $\mathcal{G}$. The requirement that the fields be single valued on spacetime is the analog of the requirement for particles that the paths do not double back in time. This quantum field theory thus posseses an equivalent Hamiltonian quantum mechanics with the preferred time directions being those of the background $\mathcal{G}$.

When the remaining sum over $\mathcal{G}$ is carried out the equivalence with any Hamiltonian formulation disappears. There is no longer any fixed geometry, any external time, to define the preferred time of a Hamiltonian formulation. All that is summed over. There would be an *approximate* equivalence with a Hamiltonian formulation were the initial conditions to imply that for large scale questions in the late universe, only a single geometry $\hat{\mathcal{G}}$ (or an incoherent sum of geometries) contributes to the sum over geometries. For then,

$$\sum_{\mathcal{G}} \sum_{\phi(x)} \exp\left(iS[\mathcal{G}, \phi(x)]\right) \approx \sum_{\phi(x)} \exp\left(iS[\hat{\mathcal{G}}, \phi(x)]\right) \tag{5}$$

and the preferred time can be that of the geometry $\hat{\mathcal{G}}$. It would be in this way that the familiar formulation of quantum mechanics emerges as an approximation to a more general sum over histories framework appropiate to specific initial conditions and our special position in the universe so far from the big bang and the centers of black holes. The preferred time of familiar quantum mechanics would then be another example of excess baggage.

In declaring the preferred time of familiar quantum mechanics excess baggage, one is also giving up as fundamental a number of treasured notions with which it is closely associated. These include the notion of causality, any notion of state at a moment of time, and any notion of unitary evolution of such states. Each of these ideas requires a background spacetime for its definition and in a quantum theory *of* spacetime there is none to be had. Such ideas, however, would have an approximate validity in the late universe on those scales where spacetime is classical. Many will regard discarding these notions as too radical a step but I think it no less radical than discarding spacetime – one of the most powerful organizing features of our experience.

6 FOR THE FUTURE

In this discussion of the excess baggage which has been discarded to arrive at a quantum mechanical framework general enough to apply to the universe as a whole, we have progressed from historical examples, through reasonable generalizations, to topics of current research and debate. Even in the last case, I was able to indicate to you candidates for the necessary generalizations. I would now like to turn to features of the theoretical framework where this is not the case; where there may be excess baggage but for which I have no clear candidates for the generalizations.

6.1 "Initial" conditions

A candidate for excess baggage is the idea that there is anything "initial" about the conditions which are needed in addition to the laws of dynamics to make predictions in the universe. Our idea that conditions are inital comes from big bang cosmology in which the universe has a history with a simple beginning ten and some billion years ago. The global picture of spacetime is that there is a more or less uniform expansion everywhere from this moment. Perhaps the very large scale structure of spacetime is very different. Perhaps, as A. Linde has suggested,[23] there are many beginnings. Our local expanding universe might be just one inflationary bubble out of many others. Perhaps, therefore, the idea that conditions should reflect a simple state at one end of the universe is excess baggage arising from our confinement to one of these bubbles. Conditions in addition to dynamics would still be needed for prediction but they might be of a very different character.[24]

6.2 Conditions and dynamics

In the framework we have discussed so far there are three kinds of information necessary for prediction in quantum cosmology: First, there are the laws of dynamics summarized by the action. Second, there is the specification of initial (or other) conditions. Third, there are our specific observations. Is it fundamental that the process of prediction be structured in this way?

Our focus on laboratory science, I believe, is the origin of the idea that there is a clean seperation between dynamics and conditions. Dynamics are governed by the laws of nature; that which we are seeking, not that which we control. By contrast, the conditions represented in the experimental arrangement are up to us and therefore not part of the laws of nature. This ideal of control is not truly realized in practice in the laboratory, but in cosmology it is never realized. The initial conditions of the universe are most definitely not up to us but must be specified in the theoretical framework in the same lawlike way as the dynamics. A law of initial conditions has, therefore, become one of the central objectives of quantum cosmology.[26] However, is it not possible that the distinction between conditions and dynamics is excess baggage arising from the focus on laboratory science to which our limited resources have restricted us? Is it not possible that there are some more general principles in a more general framework which determine both the conditions and dynamics? Is it not possible, as it were, that there is a principle which fixes both the state and the Hamiltonian?

Recent work by Hawking,[27] Coleman,[28] Giddings and Strominger[29] among others gives some encouragement to this point of view. They show that in a quantum theory of spacetime which includes wormholes — small handles on spacetime connecting perhaps widely seperated spacetime regions – that there is a closer connection between initial conditions and dynamics than had been previously thought. Specifically, the initial conditions determine the form of the Hamiltonian that we see at energies below those on which spacetime has quantum fluctuations. Indeed, they may determine a range of possibilities for the Hamiltonian only one of which is realized in our large scale universe. Dynamics and initial conditions are thereby entwined at a fundamental level.

6.3 The laws of physics

The last candidate for excess baggage that I want to discuss is the idea that the laws of physics, and in particular laws of initial conditions, are unique, apart from the universe, apart from the process of their construction, and apart from us.[30,31] Scientists, like mathematicians, proceed as though the truths of their subjects had an independent existence. We speak, for example, of "discovering" the laws of nature as though there were a single set of rules by which the universe is run with an actuality apart from this world they govern.

Most honestly, the laws of physics are properties of the collective data that we have about the universe. In the language of complexity theory, this data is compressible. There are computational algorithms by which the data can be stored in a shorter message. Take, for example, an observed history of motion of a system of classical particles. This history could be described to a given accuracy by giving the position and momentum of each particle at a suitably refined series of times. However, this message can be compressed to a statement of the positions and momenta at *one* time plus the equations of motion. Some of these initial values might be specified by observation but most, for a large system, will be specified by theory and statistically at that. For large systems the result is a much shorter message. Further, we find for many different systems that the data can be compressed to the form of initial conditions plus the *same* equations of motion. It is the universal character of this extra information beyond the initial conditions which gives the equations of motion their lawlike character. Similarly in quantum mechanics. Thus we have:

$$\text{All Observations} \rightarrow \left\{ \begin{array}{l} \text{Some Observations} \\ + \\ \text{Laws of Dynamics} \\ + \\ \text{Laws of Initial Conditions} \end{array} \right\} \text{Theory}$$

The laws of physics, therefore, do not exist independently of the data. They are properties of our data much like "random" or "computable" is a property of a number although probably (as we shall see below) in a less precise sense.[32]

This characterization of compression is incomplete. Any non-physicist knows that one can't just take a list of numbers and compress them in this way. One has to take a few courses in physics and mathematics to know what the symbols mean and how to compute with them. That is, to the list on the right should be appended the algorithms for numerical computation for practical implementation of the theory, that part of mathematics which needed to interpret the results, the rules of the language, etc. etc. Yet more properly we should write:

$$\text{All Observations} \rightarrow \left\{ \begin{array}{l} \text{Some Observations} \\ + \\ \text{Laws of Dynamics} \\ + \\ \text{Laws of Initial Conditions} \\ + \\ \text{Algorithms for Calculation} \\ + \\ \text{Mathematics} \\ + \\ \text{Language} \\ + \\ \text{Culture} \\ + \\ \text{Specific Evolutionary History} \\ + \ \cdots \end{array} \right.$$

What confidence do we have that our data can always be compressed in this way? I have discussed the possibility that a clear division between initial conditions and equations of motion may be excess baggage. Bob Geroch and I have discussed similar questions for the division between algorithms and the rest of physical theory.[33]

However it is subdivided, what guarantee do we have that the resulting theory will be unique, independent of its process of construction, independent of the specific data we have acquired? Very little it seems. As we move down the list on the right we encounter more and more items which seem particular to our specific history as a collectivity of observers and to our specific data. It is an elementary observation that there are always many theories which will fit a given set of data just as there are many curves which will interpolate between a finite set of points. Further, when the data is probabilistic as it is in quantum theory, there is always the possibility of arriving at different theories whatever criteria are used to distinguish them.

Beyond this, however, what confidence do we have that different groups of observers, with different histories, with growing but different sets of specific data, will, in the fullness of time, arrive at the same fundamental theory? I do not mean to suggest that the theories might vary because they are *consequences* of specific history, for that is not science. Rather the question is whether there are different theories which are observationally indistinguishable because of the process of their discovery.

There may be specifically cosmological reasons to expect non-uniqueness in theories of initial conditions. A theory of initial conditions, for example, must be simple enough that it can be stored within the universe. If the initial conditions amounted to some particular complex specification of the state of all matter this would not be possible. The act of constructing theories may limit our ability to find them. The gravitational effect of moving a gram of matter on Sirius by one centimeter in an unknown direction is enough to ruin our ability to extrapolate classically the motions of a mole of gas particles beyond a few seconds into the past.[34] In view of this there must be many theories of initial conditions rendered indistinguishable simply by our act of constructing them.

It would be interesting, I think, to have a framework which dispensed with the excess baggage that the laws of physics were separate from our observations of the universe, a framework in which the inductive process of constructing laws about the universe was described in it, and in which our theories were seen as but one possibility among many.

7 CONCLUSION

The assignment of the organizers was not just to speak on the subject "Where are our efforts leading?". They also wanted to know "In fifty or one hundred years time how do you think today's efforts will appear?". I have been bold enough to try their first question, I shall not be foolish enough to essay the second. I shall, however, offer a hope, and that is this: That in the future this might be seen a the time when scientists began to take seriously the idea that it was important to consider the universe as a whole and science as a unity, the time when they began to take seriously the search for a law of how the universe started, began to work out its implications for science generally, and began to discard the remainder of our excess baggage.

8 ACKNOWLEDGEMENTS

The author is grateful for the support of the John Simon Guggenheim Foundation and the National Science Foundation (NSF grant PHY 85-06686) during the preparation of this essay. He would also like to thank the Department of Applied Mathematics and Theoretical Physics, Cambridge University for hospitality while it was being written.

9 REFERENCES

1. M. Gell-Mann, Talk given at the Symposium in memory of Arthur M. Sackler to celebrate the opening of the Arthur M. Sackler Gallery at the Smithsonian Institution, September 11, 1987.

2. A number of developments in physics are considered from this point of view by the philosopher C. A. Hooker in a lecture to be published in *Bell's Theorem, Quantum Theory, and Conceptions of the Universe*, ed. by M. Kafatos (Kluwer, Boston).

3. See especially the essays "The Unity of Knowledge" and "Atoms and Human Knowledge" reprinted in N. Bohr, *Atomic Physics and Human Knowledge*, (John Wiley, New York, 1958).

4. L. Landau and E. Lifshitz, *Quantum Mechanics*, (Pergamon, London, 1958).

5. For clear statements of this point of view see F. London and E. Bauer, *La théorie de l'observation en mécanique quantique* (Hermann, Paris, 1939); R.B. Peierls, in *Symposium on the Foundations of Modern Physics*, ed. by P. Lahti and P. Mittelstaedt (World Scientific, Singapore, 1985).

6. Some have seen this division as motivation for restricting the domain of application of quantum mechanics and introducing new non-linear laws which apply in a broader domain. Examples are Wigner's discussion of consciousness [E.P. Wigner in *The Scientist Speculates*, ed. by I.J. Good, (Basic Books, New York, 1962)] and Penrose's discussion of gravitation [R. Penrose, in *Quantum Concepts of Space and Time*, ed. by C. Isham & R. Penrose, (Oxford University Press, Oxford, 1986)].

7. The original paper is H. Everett, *Rev. Mod. Phys.* **29**, 454, (1957). There is a useful collection of papers developing the Everett approach in B. DeWitt and N. Graham, *The Many Worlds Interpretation of Quantum Mechanics* (Princeton University Press, Princeton, 1973). A lucid exposition is in R. Geroch, *Noûs* **18**, 617, (1984). Even a casual inspection of these references reveals that there is considerable diversity among the various Everett points of view.

8. H. Zeh, *Found. Phys.* **1**, 69, (1971) and especially E. Joos and H.D. Zeh, *Zeit. Phys.* B **59**, 223, (1985).

9. W. Zurek, *Phys. Rev.* D **24**, 1516, (1981); *Phys. Rev.* D **26**, 1862, (1982).

10. R. Griffiths, *J. Stat. Phys.* **36**, 219, (1984).

11. See, *e.g.*, the discussion in A. Pais, *Subtle is The Lord*, (Oxford University Press, Oxford, 1982).

12. Arrows of time in quantum cosmology are discussed in S.W. Hawking, *Phys. Rev.* D **32**, 2989, (1985); D. Page, *Phys. Rev.* D **32**, 2496, (1985); S.W. Hawking, *New Scientist* July 9, 1987, p 46.

13. The arrow of time in quantum mechanics is discussed in many places. See, for example, Y. Aharonov, P. Bergmann and J.L. Lebovitz *Phys. Rev.* B **134**, 1410, (1964) and R. Penrose in *General Relativity: An Einstein Centeneray Survey*, ed. by S.W. Hawking and W. Israel (Cambridge University Press, Cambridge, 1979).

14. J.B. Hartle, "The Arrow of Time in Quantum Mechanics" (to be published).

15. This was stressed by W. Unruh in *New Techniques and Ideas in Quantum Measurement Theory*, Ann. N.Y. Acad. Sci. **480** ed. by D.M. Greenberger (New York Academy of Sciences, New York, 1986).

16. For reviews see J.A. Wheeler, in *Problemi dei fondamenti della fisica, Scuola internazionale di fisica "Enrico Fermi"*, Corso 52, ed. by G. Toraldo di Francia (North Holland, Amsterdam, 1979); K. Kuchař, in *Quantum Gravity 2*, ed. by C. Isham, R. Penrose and D. Sciama (Clarendon Press, Oxford, 1981), p. 329 ff, For recent views see the articles and discussion in Ref. 17.

17. *Proceedings of the Osgood Hill Conference on the Conceptual Problems of Quantum Gravity*, ed. by A. Ashtekar and J. Stachel (Birkhauser, Boston, 1989).

18. See, especially articles by K. Kuchař and A. Ashtekar in Ref. 17.

19. See, *e.g.*, W. Unruh and R. Wald "Time and the Interpretation of Canonical Quantum Gravity", (to be published); C. Teitelboim (to be published).

20. C. Teitelboim, *Phys. Rev.* D **25**, 3159, (1983); *Phys. Rev.* D **28**, 297, (1983); *Phys. Rev.* D **28**, 310, (1983).

21. R. Sorkin, in *History of Modern Gauge Theories*, ed. by M. Dresden & A. Rosenblum, (Plenum Press, New York, 1989).

22. J.B. Hartle, in *Gravitation in Astrophysics*, ed. by J. B. Hartle and B. Carter (Plenum Press, New York, 1987), *Phys. Rev.* D **37**, 2818, (1988); *Phys. Rev.* D **38**, 2985, (1988), (to be published) and in Ref. 17.

23. See, *e.g.*, A. Linde, *Mod. Phys. Lett.* A **1**, 81, (1986); *Physica Scripta* **T15**, 169, (1987).

24. Few of the proposals currently under investigation as candidates for the conditions of the universe genuinely have the character of an "initial" condition in the sense of posing conditions at a definite moment of time. Indeed, characteristically, as described in Section V, the notion of a single time breaks down in the early universe. Rather these proposals fix something like the quantum state of the universe for all times. But also typically these proposals imply simplicity in the early universe and a global spatial structure which consists of a single inflationary bubble. The "no boundary" proposal (Ref. 25) is a case in point. The point of the above remarks in section 6.1 is that there may be other possibilities.

25. S.W. Hawking, in *Astrophysical Cosmology*, ed. by H.A. Brück, G.V. Coyne and M.S. Longair (Pontifica Academia Scientarum, Vatican City, 1982); S.W. Hawking, *Nucl. Phys.* B **264**, 185, (1984).

26. For reviews of these efforts from the author's perspective see J.B. Hartle in *Proceedings of the International Conference on General Relativity and Cosmology, Goa India, 1987* (Cambridge University Press, Cambridge, 1989), and in *Proceedings of the 12th International Conference on General Gravitation, Boulder, Colorado, 1989* (to be published). For a bibliography of papers on this subject see J.J. Halliwell (to be published).

27. S.W. Hawking, *Phys. Lett.* B **195**, 337, (1985).

28. S. Coleman, *Nucl. Phys.* B **310**, 643, (1988).

29. S. Giddings and A. Strominger, *Nucl. Phys.* B **307**, 854, (1988).

30. J.A. Wheeler has for many years stressed the "mutability" of the laws of physics from a viewpoint which would entail much more fundamental revisions in physics than those suggested here but yet which has something in common

with them. See, *e.g.,* J.A. Wheeler in *Foundational Problems in the Special Sciences*, ed. by R. E. Butts and K. J. Hintikka (D. Reidel, Dordrecht, 1977) and J.A. Wheeler, *IBM Jour. Res. Dev.* **32**, 4, (1988).

31. For a general discussion of the status of the laws of physics and references to other points of view see, P.C.W. Davies "What are the Laws of Nature" in *The Reality Club No. 2*, ed by J. Brockman (Lynx Communications, New York, 1989).

32. For an effort at making such ideas more precise see R. Sorkin, *Int. J. Theor. Phys.* **22**, 1091, (1983).

33. R. Geroch and J.B. Hartle, *Found. Phys.* **16**, 533, (1986).

34. As discussed by H. Zeh, *Die Physik der Zeitrichtung*, Springer Lecture Notes in Physics **200**, (Springer, Heidelberg, 1984). He attributes the remark to E. Borel, *Le Hasard*, (Alcan, Paris, 1924).

THROUGH THE CLOUDS

Edward Witten
School of Natural Sciences,
Institute of Advanced Study,
Olden Lane,
Princeton, NJ 08540

When Murray Gell-Mann was starting out in physics, one of the big mysteries in the field was to understand the strong interactions, and especially the hadron resonances that proliferated in the 1950's. The existence of these resonances showed clearly that something very new was happening in physics at an energy scale of order one GeV. Another important mystery was to find the correct description of the weak interactions, and among other things to overcome the problems associated with the unrenormalizability of the simple though relatively successful Fermi theory. This problem pointed to a new development at a significantly higher energy scale.

Murray Gell-Mann played a tremendous role in advancing the understanding of these mysteries. His great contributions include his work on the strange particles; the "renormalization group" introduced by Gell-Mann and Low; early ideas about intermediate weak bosons; contributions to the proper description of the structure of the weak current; the unearthing of the SU(3) symmetry of strong interactions, and his contributions to the understanding of current algebra; the introduction of the quark model, and early ideas about QCD. His insights on these and other scientific problems are part of the foundation on which we now attempt to build, and his enthusiasm for science is an inspiration to all of us.

If we ask today what are some of the key new mysteries in particle physics, there are at least three that seem particularly pressing.

The first is the *the mystery of electroweak symmetry breaking*, which must be settled at energies at most of order 1 TeV, either by the discovery of a new structure, such as supersymmetry, or by a confirmation of the simple Higgs picture. Though the simple Higgs picture is part of the now classic SU(2) × U(1) theory in its simplest form, confirmation of this picture would be perplexing for reasons that we will come to presently.

A second problem of fundamental importance is the problem of finding the correct *unification* of forces. The energy scale that should be characteristic of this problem arises if one uses the renormalization group equation of Gell-Mann and Low to extrapolate the SU(3) × SU(2) × U(1) gauge couplings from their low energy values to an energy of possible unification at which these couplings would coincide. There are uncertainties in this process, since it depends on assumptions about how the forces fit together, and it could well be disturbed by the existence of presently unknown elementary particles. Using the known particles, and the most attractive modes of unification, Georgi, Quinn, and Weinberg, who first made this calculation nearly fifteen years ago, estimated a unification scale of order 10^{16} GeV. We could not improve on that value today. The most attractive modes of unification are those that build on the SU(5) model of Georgi and Glashow. These models give an attractive account of the quark and lepton quantum numbers. It is tempting to believe that the SU(5) structure will indeed be at least a piece of the structure of unification that will eventually emerge.

Finally, the third problem that I would like to identify is the *unrenormalizability of quantum gravity*. The energy scale characteristic of this problem is the Planck scale of 10^{19} GeV. At this energy scale, we must seriously grapple with the fact that quantum field theory and gravity seem incompatible. This fact is the one true paradox in contemporary physics, the one real analogue of the paradoxes that played such a glorious role in physics in the past. Let us recall that Maxwell theory was born because the previously known equations of electricity and magnetism were inconsistent. Special relativity was born because Maxwell theory did not make sense in the light of Newtonian mechanics. General relativity was born because Newtonian gravity did not make sense in the light of special relativity. Quantum mechanics was born in large part because a lot of things did not make sense. And the SU(2) × U(1) electroweak model was born because of the unrenormalizability of the Fermi theory. The attempt to overcome paradoxes of this nature has been one of the main inspirations for serious progress in modern physics.

The search for a proper understanding of gravity in the light of quantum mechanics can, hopefully, play a role similar to that played by analogous puzzles in the past. Certainly this search is a prime motivation for the dramatic revival of string theory that has occurred in the 1980's, with the efforts of Green, Schwarz, Brink, and later many others.

Looking at the three energy scales I have cited, it is hopefully no accident that the latter two — the energy scale of unification and the energy scale of quantum gravity — are relatively close. I say that this is hopefully no accident since our chances are making really great progress in physics in the coming epoch are much greater if this is in fact so and the Georgi-Quinn-Weinberg calculation is essentially correct.

Of course, it is quite conceivable — though to my thinking unlikely — that forthcoming experiments will reveal that the seeming rough coincidence of the unification and Planck scales really is an accident. For instance, if experiments at the TeV scale were to reveal either compositeness of quarks and leptons or a dynamical origin of weak interaction symmetry breaking, this would probably ruin grand unification as we presently understand it and mean that the apparent coincidence of the unification and Planck scales is an accident. In that case, the beautiful success of grand unification models such as SU(5) in explaining the quark and lepton quantum numbers would probably also have to be regarded as an accident. Such a turn of events, with great theoretical insights being reduced to mere accidents, would be distressing. It is also noteworthy that composite models and models of dynamical weak breaking have innumerable problems in detail. Incidentally, compositeness of quarks and leptons would come as a surprise — at least to me — at another level. The *chiral structure* of quarks and leptons is one of the fundamental observations in physics. Our understanding of this chiral structure, of course, is based on the $V - A$ structure of weak interactions that Murray Gell-Mann along with Richard Feynman, E.C.G. Sudarshan, and Robert Marshak helped unearth. This chiral structure has — it would seem — all the earmarks of something very fundamental, and there are reasons to believe that it contains information about physics at the shortest distances. It does not seem plausible, at least to me, that within the arena of quantum field theory essentially massless chiral quarks and leptons could be derived as bound states from a more satisfying starting point.

While it is hopefully no accident that the unification and gravitational energy scales are relatively close, the puzzle about the weak scale is of course that it is so much less than either of these. It is for this reason that the simple Higgs model of electroweak symmetry breaking is so problematic. This mechanism should be unstable given the existence of much greater energy scales in physics. On the other hand, we do not know of a really satisfactory alternative to the simple Higgs model. Low energy supersymmetry is very attractive in concept, and many of us dearly hope that is may be discovered. But we must realistically acknowledge that low energy supersymmetry in its known forms does not lead to such economical models as one might have hoped. I have already explained why I am even more skeptical of some of the other approaches. So we cannot claim to have a really convincing perspective about electroweak symmetry breaking. We will have to hope to learn from future accelerators whether nature has overcome the instability of the simple Higgs model, has resolved the problems of low energy supersymmetry, or has adopted another approach.

Actually, there is one major approach to the problem that I have not yet mentioned. The question of why the weak scale is so small is really a modern version of the problem that Dirac in the 1930's called the problem of the large numbers. His first "large number" was the ratio of the Planck mass to the proton mass, which is $N_1 \sim 10^{19}$. Dirac's idea was to relate this large number to another conspicuous large number in physics, namely the age of the universe in nuclear time units; this is $N_2 \sim 10^{41}$. Dirac proposed that the gravitational constant G changes in time to maintain a rough relation $N_2 \sim N_1^2$. Unfortunately, Dirac's lovely idea, one of the most beautiful suggestions that has been made about this problem, seems to be ruled out by experiment — most directly, by measurements of G/G (the rate of change time of the gravitational constant) made possible by a radio transponder landed on Mars by a Viking space craft.

Whether or not Dirac's idea can somehow be revived, the large number problem that he singled out is one off the keys to astronomy and life. The existence of complex structures in the universe, containing many elementary particles, existing for long periods, capable of storing information and giving birth to life, certainly depends among other things on the solution to the problem of large numbers. For instance, the theory of star formation shows in its most simple considerations that the number of elementary particles in a star is proportional to N_1^3. This is why the mass of the sum is about 10^{33} grams. Had nature not solved the large number problem, there would be no astronomy as we know it, and certainly no life.

The advances of the last thirty or forty years of course play a great role in how we think about the problems I have cited. In particular, nowadays one has much greater confidence in the scope and power of relativistic quantum field theory than one had when Murray Gell-Mann entered the scene. Most physicists seem to believe that whatever is discovered about electroweak symmetry breaking and the large numbers problem will have a natural explanation in terms of quantum field theory. I certainly share this view. By rights, a great, rich, and successful framework such as quantum field theory should break down only in the face of a truly formidable adversary. Gravity is the natural and almost inevitable candidate.

Another result of the advances of the last decade is that we have acquired a much sharper confidence in the role of geometry in physics — in *quantum* as well as *classical* physics. Above all this comes from what we have learned about *quantum gauge theories* as well as from investigations of *supersymmetry* and *string theory.*

In the early decades of quantum field theory, geometry seemed an isolated preserve of classical relativity. Quantum physicists had much more immediate and down to earth problems on their hands, connected with the cumbersomeness and internal difficulties of quantum field theory and its phenomenological problems. But the developing understanding of the role of non-abelian gauge theories in physics has enabled us in the last ten or fifteen years to begin to appreciate the role of geometry in quantum field theory and to begin thinking about quantum field theory in geometrical terms. This is a development whose potentially far reaching implications are just beginning to be appreciated.

Thus, our period will be remembered first of all as the epoch in which quantum field theory came of age, the scope of her physical applications vastly expanded, her vitality no longer questioned, her frailties and "anomalies" now seen as signs of grace and beauty.

It will be seen as the era when quantum field theory, thus coming of age, emerged as a truly fit partner for a fateful match with that other great current of Riemannian geometry and general relativity.

It will be seen as a time in which progress in understanding the strong, weak, and electromagnetic interactions led physicists — unwittingly at first — to become geometers again.

It will be seen as a time in which the serendipitous discovery of dual models — and the great discoveries that followed in string theory — opened gaps in the clouds, and permitted physicists to begin grappling with problems that otherwise would have been reserved for the twenty-first century.

It will be seen as a time in which — through these gaps in the clouds — physicists caught their first glimpses of a dramatic landscape that we mostly can still only dream about.

Finally, it will be seen as an era in which the geometrical tools at our disposal lag behind what we need to properly appreciate the insights glimpsed through the clouds, and in which *the search for geometric understanding* is thus likely to be one of the great engines of progress.

Covariant Formulations of the Superparticle
and
the Superstring

Lars Brink

Institute of Theoretical Physics

S-412 96 GÖTEBORG, Sweden

1 Introduction

Supersymmetry is the first real extension of space-time symmetry. It has given us great hope that we should be able to generalize ordinary geometry into a supergeometry and in this process obtain more unique and consistent models of physics. In some cases this has been achieved, but in most cases we still lack a natural and unique extention into a superspace.

The concept of superspace, i.e., a space with fermionic coordinates as well as bosonic coordinates, was introduced first in dual models by Montonen [1] in an attempt to construct multiloops in the Ramond-Neveu-Schwarz model [2]. This led eventually to the superconformal algebras and super-Riemannian spaces [3]. When supersymmetric field theories were discovered [4], it was soon realized that a superspace is the natural space in which to describe these models [5]. However, these descriptions, although in the end quite successful in establishing renormalization properties [6], always lacked a certain sense of naturalness. For each supermultiplet different ideas had to be used.

In supergravity theories [7], being extensions of truly geometric theories, the hopes were even higher and the results more discouraging. So far one has only managed to write superspace actions for the N=1 theory [8], and none of them is a natural extension of the Hilbert action. Superspace techniques were though eventually useful in describing the classical theories [9] and led to the really important result that any supergravity theory has infinitely many possible counterterms [10]. This is perhaps the most important result in supergravity theory, since it shows that we must expect these theories to diverge uncontrollably in the quantum case.

The most natural superspace arose when we realized that supersymmetric theories are streamlined to be described in the light-cone gauge [11]. Here one can eliminate all unphysical degrees of freedom and the physical ones fit nicely into one single superfield, making up all its components. This formalism was quite successful and led among other things to the proof of finiteness for the N=4 Yang-Mills theory [12]. However, the light-cone gauge is a gauge in which geometry is obscured and its virtues are more on the technical side than on the conceptual one.

Superspace techniques have become a natural framework to describe the Ramond-Neveu-Schwarz model [13] and its projection to the superstring [14]. It was also instrumental in the construction of new string models such as the SO(2) model [15] and the SU(2) one [16]. However, again when one tried to introduce superworld-sheet actions with superreparametrization invariance, only somewhat unnatural formulations were found, reflecting the same problem as in supergravity theories [17].

The superstring theory carries space-time supersymmetry and it is natural to ask if the space-time coordinates of this theory can be fit into a superspace. When attempting such a scheme, unexpected difficulties arise. In the original version one finds for example that it cannot be quantized covariantly, at least not in a straightforward way [18]. Several solutions to this dilemma have been put forward. They all involve adding new coordinates or momenta for the string or use an infinity of ghost coordinates, and none of them seems to be a very natural solution. In this report I will discuss these various attempts and compare their virtues.

It should be said that we might be studying an academic problem. In a string theory the world-sheet is really the important space, where the physics is defined, and the RNS model is built by demanding supersymmetry on this space. Space-time is somehow coming out as a solution and we must accept what we get. Furthermore we know that space-time will not really make sense in the very early universe at Planck energies, and if we look for a fundamental formulation that can be taken back to the earliest times we should certainly concentrate on the world-sheet physics. However, my hope is that a simultaneous study of the space-time and the world-sheet can lead us to the best set of variables quicker.

The perhaps most fascinating and profound discovery in string theory is the possible phase transition in the very early universe [19]. It gives us a logically consistent picture of how our universe, i.e., the string phase we are in now, was created with a finite energy density and temperature thus avoiding singularities. A truly fundamental description of physics should be able to describe both phases. However, it is most plausible that in the earlier phase, concepts such as space and time are not meaningful quantities. The really great question is to find the quantities that can describe both phases. Perhaps the problem above can shed light on this issue.

Let me finally raise one more question that at least bothers me. String theory is most naturally defined in a Euclidean space. All techniques based on Riemann surfaces are really Wick-rotated formulations. This fact was not too well understood in the days of dual models, but became clear in Polyakov's functional formulation [20]. Here it is also straightforward to compute the correct loop graphs, having factored out the modular transformations in the case of closed strings. If one tries to construct these loop graphs from covariant operator rules using unitarity (and hence a Minkowskian formulation) one obtains these loops with the modular group not divided out. (In a light-cone gauge treatment, which is inherently a Minkowskian treatment one gets the correct result.) I would like to interpret this fact to cast some shadow on the time variable.

In this report I will be somewhat brief in the detailed descriptions of the various attempts to quantize the superstring covariantly. It would take me too far to go into all details and I refer to the original papers for the details. Instead I will try to show the virtues and problems with each attempt.

2 The Minimal Covariant Action

To describe a string theory with space-time supersymmetry the natural coordinates are $x^\mu(\sigma,\tau)$ and $\theta^\alpha(\sigma,\tau)$, where x^μ and θ^α are vectors and spinors resp. under SO(1,d-1) with d being the dimension of space-time, which we take to be 10. The momentum density

$$\pi_\alpha^\mu = \partial_\alpha x^\mu - i\bar\theta\gamma^\mu\partial_\alpha\theta \quad (\alpha = (\tau,\sigma)) \tag{2.1}$$

is the natural supersymmetrically invariant extension of the momentum density used for bosonic strings. To construct an action the natural thing is to insert π_α^μ instead of $\partial_\alpha x^\mu$ in the action for the bosonic string. This is, however, not enough. To obtain a local fermionic symmetry which can eliminate unphysical fermionic degrees of freedom Green and Schwarz added an extra (invariant) term, a Wess-Zumino-Novikov-Witten term in the language of σ-models and suggested the action [21]

$$S = -\frac{1}{2}\int d\tau d\sigma \left[\sqrt{-g}g^{\alpha\beta}\pi_\alpha\cdot\pi_\beta - 2i\epsilon^{\alpha\beta}\partial_\alpha x^\mu\bar\theta\gamma_\mu\partial_\beta\theta\right]. \tag{2.2}$$

To understand the problems of this action it is easier and equally informative to study the point particle limit

$$S_p = -\frac{1}{2}\int d\tau\, e^{-1}\pi^\mu\pi_\mu \tag{2.3}$$

$$\pi^\mu = \dot x^\mu - i\bar\theta\gamma^\mu\dot\theta \tag{2.4}$$

and e is the einbein $\sim \sqrt{g_{\tau\tau}}$.

In a Hamiltonian formalism one gets the following primary constraints

$$p_e = 0 \tag{2.5}$$

$$\chi = \bar p_\theta + i\bar\theta\slashed{p} = 0, \tag{2.6}$$

where

$$p^\mu = -\frac{1}{e}\pi^\mu.$$

The secondary constraint is

$$p^2 = 0. \tag{2.7}$$

In order to check if the constraints correspond to gauge symmetries of the action we must check if the constraint algebra, obtained by using the canonical Poisson brackets, closes. The critical bracket in the algebra turns out to be

$$\{\bar\chi_\alpha,\bar\chi_\beta\} = 2i(\gamma_0\slashed{p})_{\alpha\beta}, \tag{2.8}$$

where the RHS clearly is not a constraint. The rank of this 16×16 matrix is 8 because of the constraint (2.7) and this fact shows that 8 of the 16 constraints $\bar{\chi}$ are not gauge constraints, i.e., they are second class constraints in Dirac's terminology and should be eliminated [18]. However, there is no covariant way of dividing up χ into two 8-component spinors. It is true that $\displaystyle{\not}p\chi$ is effectively an 8-component spinor because of (2.7), but there is no other vector satisfying (2.7) in the theory that can be used to project out the other 8-component spinor. This is the root of the problem. If we allow ourselves to break covariance there is no problem, and we can easily quantize the system in the light-cone gauge where only SO(8) covariance is maintained, since $16=8_s+8_c$ under the decomposition SO(1.9)→SO(8). (The representations 8_s and 8_c are the two 8-dimensional spinor representations of SO(8).)

Various methods have been developed to treat systems with second-class constraints in the BRST-treatment. In the case above one can start by constructing a BRST-charge Q_B including all 16 constraints, hence introducing 16 ghost coordinates of bosonic type. This is an overcounting and must be compensated by a new set of 16 ghosts for ghosts, which in turn must be compensated by 16 ghosts for ghosts for ghosts and this procedure goes on ad infinitum. There is, in fact, a problem even with the set of constraints $p^2 = 0$ and $\displaystyle{\not}p\chi = 0$, since they are not independent of each other. A proper BRST formulation needs an infinity of ghosts. This fact shows up in a Lagrangian formulation by the effect that the gauge symmetry related to the constraint (2.6) (the κ-symmetry) does not close off-shell. Also here one can show that it leads to an infinity of ghosts [22]. Gauge-fixed actions which are quadratic in the infinity of coordinates (including the ghosts) have been constructed but it is still unclear if they properly take the second-class constraints into account.

We could, in fact, have been suspicious from the beginning. From supersymmetry we conclude that P^o is a positive operator in the quantum case, i.e., there are no negative energy states. However, a superstring clearly contains spinning states that for covariant descriptions need negative energy states. A rigorous deduction of the statement that $P^o > 0$ requires a proper time gauge quantization in a positive definite Hilbert space, which has not been done. However, in retrospect of the result above we can certainly trust it and use the reasoning above as a reference for attempts to quantize covariantly [23]. It should be mentioned that the obstructions above does not apply in the light-cone gauge, since all states have positive energy in light-cone variables.

Perhaps the most important aspect here is the theorem of Jordan and Mukunda [24], which states that no covariant commuting position operators for spinning particles can be defined on a Hilbert space spanned by positive energy states only. In quantizations of systems with only first class constraints the position operators are certainly commuting and the theorem above signals second-class constraints. In fact, by quantizing (2.3) in the light-cone gauge using the Dirac procedure we arrive at

$$[x^\mu, x^\nu] = \frac{-p_\rho}{2p^2}\bar{\theta}\gamma^{\mu\nu\rho}\theta .\qquad(2.9)$$

Since covariance is already broken we can define a new position operator [25]

$$q^\mu = x^\mu + \frac{ip_\nu}{2p^+}\bar{\theta}\gamma^{\mu\nu+}\theta ,\qquad(2.10)$$

which does commute with itself and is canonically conjugate to p^μ.

If we are to use fields which are functions of positions we certainly need commuting position operators and the reasoning above must be kept in mind.

3 Covariantized Light-Cone Approach

There is a rather straightforward way to covariantly eliminate the second-class constraints by making the light-cone decomposition not with respect to a given frame but rather with respect to two null vectors which are not fixed a priori and are treated as dynamical variables [26]. More precisely, we introduce two null vectors n^μ and r^μ obeying

$$\begin{aligned} n^2 &= 0 \\ r^2 &= 0 \\ r\cdot n &= -1 . \end{aligned}\qquad(3.1)$$

The pair (n^μ, r^μ) can now replace the $(+,-)$ directions of the light-cone analysis.

In order to leave the dynamical content of the original system unchanged the new variables should be pure gauge. In the Hamiltonian formalism we then also impose the constraints that the momenta π_μ and σ_μ conjugate to n^μ and r^μ vanish.

$$\pi_\mu = \sigma_\mu = 0\qquad(3.2)$$

The variables (n^μ, r^μ), subject to (3.1) parametrize the coset space $SO(1,9)/SO(8)$ since the stability group is isomorphic to $SO(8)$ rotations.

The whole system is now described by constraints (2.5)-(2.7) and (3.1)-(3.2) together with the naive canonical Poisson brackets among the variables. There are new second-class constraints among the constraints (3.1)-(3.2) which further complicate the final Dirac brackets and hence the commutators. However, it is now straightforward to decompose the constraints $\bar{\chi}$ (2.6) into the second-class piece and the first-class one in a covariant manner. Indeed one can identify the second-class constraints as

$$\bar{\psi} = \bar{\chi}\slashed{n}\slashed{r}\qquad(3.3)$$

while the first class constraints are

$$\overline{\varphi} = \overline{\chi}\not{p} \qquad (3.4)$$

According to the theorem we noted in the last section we still must expect to get a non-commuting position operator. Indeed we get

$$[x^\mu, x^\mu] = \frac{1}{4 n \cdot p} \overline{\theta}\gamma^\mu \not{n} \gamma^\nu \theta \qquad (3.5)$$

As in the light-cone gauge one can though find a shifted position variable canonically conjugate to p_μ which is commuting.

$$q^\mu = x^\mu + \frac{i}{2} \frac{1}{n \cdot p} n_\rho p_\sigma \overline{\theta}\gamma^{\mu\rho\sigma}\theta - \frac{n^\mu}{2(n \cdot p)^2} \overline{p}_\theta \not{p} \theta \qquad (3.6)$$

Note that we have now evaded the theorem. The price we have had to pay is an increased number of variables. This is a heavy price, since it means that in a wave function representation or a field theory the functions will be functions of an increased number of (bosonic) variables. It also turns out that this extension of variables although minimal in a certain sense leads to fairly complicated commutators among the extra variables making the wave function representation somewhat obscure and the usefulness of the method doubtful.

4 Harmonic Superspace

We have seen in the preceding sections that light-cone gauge covariance (SO(8)) is very easily achieved in the quantization of the superstring, while the full one (SO(1,9)) is harder. This is very much reminiscent of the situation in supersymmetry, where simple N=1 supersymmetry is easy to implement, while the higher N supersymmetries are much more difficult. A way out of this dilemma was devised by Ogievetsky and his collaborators [27]. They reduce a global symmetry group G to a subgroup H by introducing harmonic variables u_H^G transforming covariantly under both G and H. By building in supersymmetry covariantly in H one can use the harmonic variables to implement the G symmetry covariantly. The price one has paid is the introduction of new fields u. The formalism is really borrowed from the vierbein formalism of gravity where the vierbein fields V_μ^m transform between G=GL(4,R) and H=SO(1,3) (for d=4).

A similar programme for the superstring has been carried out by Nissimov, Pacheva and Solomon [28]. They introduce new bosonic coordinates

$v_\alpha^{\pm 1/2}$ 2 Majorana-Weyl, spinors
u_μ^a 8 Lorentz vectors.

which satisfy the kinematical constraints.

$$\left(\bar{v}^{+\frac{1}{2}}\gamma^\mu v^{+\frac{1}{2}}\right)\left(\bar{v}^{-\frac{1}{2}}\gamma_\mu v^{-\frac{1}{2}}\right) = -1, \tag{4.1}$$

$$\left(\bar{v}^{\pm\frac{1}{2}}\gamma^\mu v^{\pm\frac{1}{2}}\right) u_\mu^a = 0, \tag{4.2}$$

$$u_\mu^a u^{b\mu} = \delta^{ab}. \tag{4.3}$$

The group SO(8) x SO(1,1) act on $u_\mu^a, v_\alpha^{\pm 1/2}$ as an internal group of local rotations with u_μ^a transforming as a vector under SO(8), (it could also be chosen as a spinor), and $v^{\pm 1/2}$ carry the charge $\pm 1/2$ under SO(1,1). Through a famous Fierz identity one can easily prove that the composite Lorentz vectors

$$u_\mu^\pm = \bar{v}^{\pm\frac{1}{2}}\gamma_\mu v^{\pm\frac{1}{2}} \tag{4.4}$$

are light-like. With the help of $u_\mu^\pm$ and u_μ^a it is straightforward to disentangle Lorentz vectors and tensors covariantly to "light-cone components"

$$A^a \equiv u_\mu^a A^\mu; \quad A^\pm = u_\mu^\pm A^\mu \tag{4.5}$$

As in the case in the preceding section we must now add an action for the harmonic variables making them pure gauge. Here we should note that both in the superparticle and in the superstring case we can use harmonic variables that depend on only one parameter, τ. We can fix the same harmonic frame at each value of σ in the string case.

As in the case in the preceding section it is now straightforward to decompose a 16-component spinor into two 8-component ones. For the two chiralities of SO(1,9) one has

$$\psi_\alpha = \left(\gamma^a\gamma^+ v^{-1/2}\right)\psi_a^{-\frac{1}{2}} + \left(\gamma^a\gamma^- v^{+1/2}\right)\psi_a^{+\frac{1}{2}} \tag{4.6}$$

$$\phi^\alpha = \left(\gamma^a v^{+1/2}\right)\phi_a^{-\frac{1}{2}} + \left(\gamma^a v^{-1/2}\right)\phi_a^{+\frac{1}{2}} \tag{4.7}$$

In this way we can again decompose the constraint (2.6) covariantly into its first- and second-class parts. One advantage here over the one in sect. 3 is that no new second-class constraints are introduced from the new (harmonic) variables. In fact, the formalism even allows us to eliminate the second-class ones altogether. Note that in the Dirac quantization procedure one adds gauge fixing conditions to the first-class constraints. The total set of these conditions must then constitute second-class constraints. One can turn this procedure around. On a set of second-class constraints one can attempt to divide them up into one set which is first-class and one which is to be regarded as gauge fixing conditions. In the case of the superstring augmented with harmonic variables one can in fact perform such a divison and be left with only first-class constraints.

In the BRST-description of this system one finds that a finite number of ghosts is sufficient.

Concluding this section we have found that the introduction of harmonic variables does allow us to quantize covariantly. Again the price we have had to pay is an increase in variables that the wave functionals depend on, although in the string case the new variables are constant in σ. This method has been quite successful and Nissimov, Pacheva and Salomon have reached a number of results. The possible objections to this method and the one in sect. 3 is that they are covariantized light-cone formalisms and would contain the same amount of information as light-cone gauge formulations.

5 Actions with only first-class constraints

There is, in fact, a way to write actions in terms of the original variables only which has been pursued by Siegel [29]. Let us go back to the constraints (2.5) and (2.6). Let me use Siegel's notation d for χ. The set of constraints

$$\mathcal{A} = p^2 \qquad (5.1)$$
$$\mathcal{B} = \slashed{p}d . \qquad (5.2)$$

are clearly first-class as was noted in (3.4). This system can, however, not be complete since the second-class constraints are simply dropped. Siegel then found another set of constraints that can be added, namely

$$C^{\mu\nu\rho} = \bar{d}\gamma^{\mu\nu\rho}d \qquad (5.3)$$

The algebra (5.1)-(5.3) is closed. The new feature here is the appearance of constraints bilinear in fermionic operators, which hence cannot be solved. To compare this model with the original one in sect. 2 one can go to the light-cone gauge. The constraints (5.2) are then solved for, but the $C^{\mu\nu\rho}$'s must be treated in a BRST formulation. One finds two sections, one physical and one unphysical, and the physical one can be seen to agree with the one following from the action (2.3).

This formulation of the superparticle seems to violate the theorem of Jordan and Mukunda alluded to before. However, a detailed study shows that this is not the case [30]. Since the constraints (5.3) cannot be solved for they must be imposed on the physical states. The set of states must then be augmented with unphysical ones and one can show that among these ones there are states with negative energy. Hence the prerequisites for the theorem is avoided.

In the string case a convenient covariant gauge choice is to introduce

$$g_{\alpha\beta} = e^{\phi}\eta_{\alpha\beta} \tag{5.4}$$

and similar conditions for the possible other members of the 2-dimensional supergravity multiplet. This conformal gauge is a partial gauge fixing at the hamiltonian level and the remaining constraints are either treated in a BRST-formulation or used to project out the physical states. For the action (2.2) this procedure is hampered by the occurrence of the second-class constraints. In Siegel's approach he simply conjectures a set of constraints in the conformal gauge and then checks that it closes and that this formalism agrees with the older one in the light-cone gauge.

To start Siegel constructs operators that mimic the p^{μ}, d^{α} of the particle case.

$$P^{\mu}(\sigma) = p_x^{\mu} + x'^{\mu} + i\bar{\theta}\gamma^{\mu}\theta' \tag{5.5}$$

$$D^a(\sigma) = p_{\theta}^a + \gamma_{\mu}\theta\left(p_x^{\mu} + x'^{\mu}\right) + \frac{i}{2}\gamma^{\mu}\theta\bar{\theta}\gamma_{\mu}\theta' \tag{5.6}$$

$$\Omega^a(\sigma) = i\theta'^a , \tag{5.7}$$

which satisfy the Kac-Moody algebra

$$[P^{\mu}(\sigma), P^{\nu}(\sigma')] = i\delta(\sigma - \sigma')\eta^{\mu\nu} \tag{5.8}$$

$$\{D^a(\sigma), D^b(\sigma)\} = 2\delta(\sigma - \sigma')(\gamma \cdot P(\sigma))^{ab} \tag{5.9}$$

$$[D^a(\sigma), P^{\mu}(\sigma')] = 2\delta(\sigma - \sigma')(\gamma^{\mu}\Omega(\sigma))^a \tag{5.10}$$

$$\{D^a(\sigma), \Omega^b(\sigma')\} = i\delta'(\sigma - \sigma')\delta^{ab} , \tag{5.11}$$

the rest being zero.

From these operators one can construct a superconformal algebra with the generators

$$A = \frac{1}{2}P^2 + \bar{\Omega}D \tag{5.12}$$

$$B^a = (\gamma \cdot PD)^a \tag{5.13}$$

$$C^{ab} = \frac{1}{2}\bar{D}^{[a}D^{b]} \tag{5.14}$$

$$D^{\mu} = i\bar{D}\gamma^{\mu}D . \tag{5.15}$$

Through a lenghty calculation one can check that these generators close into an algebra, where the structure coefficients are field dependent. By going to the light-cone frame again one finds agreement with the standard superstring.

The algebra among the generators (5.11)-(5.14) is an extension of the Virasoro algebra. It is outside the standard classification of super-Virasoro algebras, in fact making up an N=16 algebra. The reason why it can work is the appearance of field dependent structure coefficients. Such occurrencies usually signal a partial gauge fixing. It is an intriguing question whether the algebra can be further extended to avoid the field dependence. A further complication is the fact that the generators (5.11)-(5.14) are not independent of each other. To find an independent set one must divide up the generators in light-cone components thus ruining the covariance.

The non-independence also complicates the BRST procedure. Again one is forced to introduce ghost for ghosts ad infinitum and such a formalism ought to coincide with the one from the original action (2.2).

The formalism described in this section is somewhat more general than the one in sect. 2. It allows in principle for a covariant quantization. In the case of the point-particle it does give an explanation why covariant superfield methods are working although it does not give a unified and natural framework for all supermultiplets as mentioned in the introduction. In the superstring case the advantages with this formalism is probably negligable. One could attempt an old-fashioned operator formalism without ghosts but it would be fairly awkward because of the complicated super-Virasoro algebra. In fact in such an attempt I see no clear way to implement unitarity. Introducing the ghosts, which I think is necessary would lead us back to the formalism in sect. 2.

6 Twistors

The formalism described so far have been rather conventional in the sense that they aim at describing superstring theory in terms of space-time coordinates augmented with other coordinates. However, if we are really trying to find variables that will be fundamental and can describe both phases of superstring theory we should search for variables from which space-time could be derived. We have very little guidance here. If the Poincaré group is the underlying symmetry group there are though a rather limited number of alternatives. (We may have to be more imaginative!) One such proposal which have been put forward by Bengtsson, Bengtsson, Cederwall and Linden [31] and originally by Shirafuji [32] is to use twistor variables.

There are two properties among twistors that are appealing in this connection.
(i) Twistors do substitute for x^μ and p^μ.
(ii) There is a close connection between twistors and division algebras and the dimensions in which superstrings can be described.

The basic relations are the local isomorphisms between the Lorentz groups $SO(1,\nu+1)$

and the groups $SL(2, \vec{K}_\nu)$, where

$$\vec{K}_\nu = \vec{R}, \vec{C}, \vec{H}, \vec{O} \text{ for } \nu = 1, 2, 4 \text{ and } 8,$$

the four division algebras. This means that for d=3,4,6 and 10 twistor formulations are possible and these are the dimensions in which the classical superstring can exist. It should be said here that d=10 is more complicated than the others since the octonionic algebra is non-associative. In the sequel I will for simplicity only discuss d=3, 4 and 6.

The starting-point now is to use the isomorphism above to write every Lorentz vector as a bispinor under $SL(2, \vec{K}_\nu)$

$$V^\mu \to \vec{V}^{\alpha\dot\alpha} = \begin{pmatrix} \sqrt{2}V^+ & \vec{V} \\ \vec{V} & \sqrt{2}V^- \end{pmatrix}$$

$$\vec{V} \in \vec{K}_\nu$$

(6.1)

$$\vec{V} = \sum_{L=1}^{\nu} V_i e_i \tag{6.2}$$

with e_i an orthonormal basis for $\vec{K}$.

In the point-particle case the free bosonic particle is described by the action

$$S = \int d\tau \left[\frac{1}{2} Sc \left(\dot{\vec{x}}^{\alpha\dot\alpha} \vec{p}_{\alpha\dot\alpha} \right) + V \det \vec{p} \right] \tag{6.3}$$

where Sc means scalar part under the division algebra and V is an auxiliary field (determinant of the einbein, if we so wish). We immediately get the constraint

$$\det \vec{p} = 0, \tag{6.4}$$

which means

$$p_\mu p^\mu = 0. \tag{6.5}$$

This can be solved by the constraint

$$\vec{\pi}_{\alpha\dot\alpha} = \vec{p}_{\alpha\dot\alpha} - \vec{\psi}_\alpha \vec{\psi}_{\dot\alpha} = 0, \tag{6.6}$$

where we introduce the bosonic spinor (twistor) $\vec{\psi}_\alpha$ and its complex conjugate. The constraint (6.5) is automatic because of the famous Fierz identity, which we used in (4.4). We furthermore introduce the bosonic spinor $\vec{\omega}^\alpha$ according to

$$\vec{g}^\alpha = \vec{\omega}^\alpha - \vec{\psi}_{\dot\alpha} \vec{x}^{\alpha\dot\alpha} = 0 \tag{6.7}$$

We can now use ω and ψ as canonical variables instead of x and p with the Poisson bracket

$$\{\omega_i^\alpha, \psi_{j,\beta}\} = \delta_{ij}\delta_\beta^\alpha \qquad (6.8)$$

Furthermore we can write an action which contains both sets

$$S = \int d\tau\, Sc\left\{\frac{1}{2}\vec{x}^{\dot\alpha\alpha}\vec{p}_{\alpha\dot\alpha} + \vec{\omega}^\alpha\dot{\vec\psi}_\alpha + \frac{1}{2}T^{\alpha\dot\alpha}\vec\pi_{\alpha\dot\alpha}(\vec p,\vec\psi) + \vec\gamma_\alpha\vec g^\alpha(\vec\Omega,\vec\psi,\vec x)\right\} \qquad (6.9)$$

It contains enough gauge invariance to allow for a gauge $\vec\omega = \vec\omega_f$, $\vec\psi = \vec\psi_f$ and obtain the usual description in terms of x and p, or $\vec x = \vec x_f$, $\vec p = \vec p_f$ and obtain the twistor formulation.

In the twistor formulation the mass-shell condition $P^2 = 0$ is built in by construction. Instead the on-shell condition reads

$$u = \frac{1}{2}\left(\vec\omega^\alpha\vec\psi_\alpha - \vec\psi_\alpha\vec\omega^\alpha\right) = 0 \qquad (6.10)$$

For d=4 this is the spin-shell condition. The twistor formulation hence reverses the order in which the Casmir invariants are treated. Off-shell means that the helicity is continuously varied. This fact means that interactions would be very much different in such a formulation and it is so far unclear if one can describe interactions in this language.

The twistor formalism is quite suitable also in the superparticle case and it lends itself naturally to a covariant quantization. As in the bosonic case one can start with a master action containing both the ordinary space-time coordinates and the twistors

$$S = \int d\tau\, Sc\left[\frac{1}{2}\pi^{\dot\alpha\alpha}\vec p_{\alpha\dot\alpha} + \vec\omega^\alpha\dot{\vec\psi}_\alpha + \frac{i}{2}\xi\dot{\vec\xi} + \frac{1}{2}T^{\alpha\dot\alpha}\vec\pi_{\alpha\dot\alpha} + \vec\gamma_\alpha\vec g^\alpha + \vec{\bar\rho}\vec r\right], \qquad (6.11)$$

where

$$\pi^{\dot\alpha\alpha} = \dot{\vec x}^{\dot\alpha\alpha} - i\left(\dot\theta^{\dot\alpha}\bar\theta^\alpha - \theta^\alpha\dot{\bar\theta}^{\dot\alpha}\right), \qquad (6.12)$$

where

$$\vec\pi_{\alpha\dot\alpha} \equiv \vec p_{\alpha\dot\alpha} - \vec\psi_\alpha\vec\psi_{\dot\alpha}, \qquad (6.13)$$

$$\vec g^\alpha \equiv \vec\omega^\alpha - \vec\psi_\alpha\left(\vec x^{\dot\alpha\alpha} - i\vec\theta^\alpha\bar\theta^{\dot\alpha}\right), \qquad (6.14)$$

$$\vec r = \vec\xi - \sqrt{2}\vec\psi_\alpha\theta^\alpha. \qquad (6.15)$$

Gauge invariance implies that we can either choose a gauge described by x and θ or one in terms of the twistor variables ω, ψ and ξ.

$$\{\omega_i^\alpha, \psi_{j\beta}\} = \delta_{ij}\,\delta_\beta^\alpha \qquad (6.16)$$
$$\{\xi_i, \xi_j\} = -i\delta_{ij} \qquad (6.17)$$

and one finds straightforwardly that there are only first class constraints. It is now fairly direct to quantize and to introduce wave functions (and fields) with twistors as coordinates. It is, as said above, a challenging task to try to write an interacting field theory this way.

The programme has so far not been fully implemented to the octonionic case (d=10), but there is good hope that it can be done. For strings there are further problems to overcome. Here one has two light-like vectors $\partial_+ x^\mu$ and $\partial_- x^\mu$ (according to the Virasoro conditions). The first guess would be to introduce two different twistors, one for each vector. However, it sounds like an overcounting to introduce two twistors for a vector. Some progress on this problem has recently been done by Cederwall [33].

The twistor approach is a challenging idea, which should be investigated to see if it can be used to describe interacting superstrings. It has the virtue of using a minimal set of variables out of which space-time can be constructed. However, it is a formalism in which we use representations of the Lorentz group and it is not clear to me that the original phase of superstrings need be described by that symmetry. In fact, there are lots of reasons to believe that the symmetry in the original phase is much bigger.

7 Conclusions

Quite a number of methods have been devised in order to implement a covariant quantization of the superstring. Apart from the twistor approach, all the methods use an enlarged phase space, either an infinite series of ghosts for ghosts or new bosonic coordinates. These formalisms are certainly going to give new insight into superstring theory. In fact they might be quite useful in one of the great issues in superstring theory, namely in the search to get an understanding of non-perturbative effects. We know that they must play a role, for example in seeking a true minimum among all the classical solutions, all conformal field theories with no anomalies. Here it is important to have as efficient a formalism as possible. I think this question justifies all the efforts made in order to find a covariant formalism.

The other, perhaps even more important and certainly deeper issue is to understand the possible phase transition in the early universe. For this issue I do not think we have the best formalism yet. It is not clear to me that we can use the standard coordinate to describe such a phase transition. In fact, we know that it occurs at Planck energies, where space-time really breaks down because of quantum fluctuations. Can we find more fundamental variables from which for lower energies space-time can be constructed? Is there a deeper level at which the phase transition can be understood from some physical principles? These are fascinating questions which we know very little about now. I think the search for answers to them is the real challenge in front of us in superstring physics.

References:

1. C. Montonen, Nuovo Cim. **19A** (1974), 69

2. P.M. Ramond, Phys. Rev. **D3** (1971), 2415
   A. Neveu and J.H. Schwarz, Nucl. Phys. **B31** (1971), 86; Nucl. Phys. **B31** (1971), 86; Phys. Rev. **D4** (1971) 1109

3. L. Brink and J.O. Winnberg, Nucl. Phys. **B103** (1976), 445
   D. Friedan, E. Martinec and S. Shenker, Nucl. Phys. **B271** (1986) 93

4. Yu. A. Golfand and E.P. Likhtman, JETP Lett. **13** (1971), 13
   J. Wess and B. Zumino, Nucl. Phys. **B70** (1974), 39

5. A. Salam and J. Strathdee, Nucl. Phys. **B76** (1974), 477

6. M.T. Grisaru, W. Siegel and M. Rocek, Nucl. Phys. **B159** (1979), 429

7. D.Z. Freedman, P. van Nieuwenhuizen and S. Ferrara, Phys. Rev. **D13** (1976), 3214
   S. Deser and B. Zumino, Phys. Lett. **62B** (1976), 335

8. J. Wess and B. Zumino, Phys. Lett. **74B** (1978), 51
   L. Brink and P. Howe, Phys. Lett. **88B** (1979), 81

9. L. Brink and P. Howe, Phys. Lett. **88B** (1979), 268

10. P.S. Howe and U. Lindström, Nucl. Phys. **B181** (1981), 487
    R.E. Kallosh, Phys. Lett. **99B** (1981), 122

11. L. Brink, O. Lindgren and B.E.W. Nilsson, Nucl. Phys. **B212** (1983), 401

12. S. Mandelstam, Nucl. Phys. **B213** (1983), 149
    L. Brink, O. Lindgren and B.E.W. Nilsson, Phys. Lett. **123B** (1983), 323

13. See the first reference in [3]

14. M.B. Green and J.H. Schwarz, Nucl. Phys. **B181** (1981), 502

15. M. Ademollo, L. Brink, A. D'Adda, R. D'Auria, E. Napolitano, S. Sciuto, E. Del Giudice, P. Di Vecchia, S. Ferrara, F. Gliozzi, R. Musto, R. Pettorini and J. Schwarz, Nucl. Phys. **B111** (1976), 77

16. M. Ademollo, L. Brink, A.D' Adda, R. D'Auria, E. Napolitano, S. Sciuto, E. Del Giudice, P. Di Vecchia, S. Ferrara, F. Gliozzi, R. Musto and R. Pettorini, Nucl. Phys. **B114** (1976), 297

17. P.S. Howe, J. Phys. A. Math. Gen. Vol. **12**, No. 3 (1979), 393

18. I. Bengtsson and M. Cederwall, ITP-Göteborg **1984-21**

19. B. Sundborg, Nucl. Phys.**B254** (1985), 583
    M.J. Bowick and L.C.R. Wijewardhana, Phys. Rev. Lett. **54** (1985), 2485

20. A.M. Polyakov, Phys. Lett. **103B** (1981), 207, 211

21. M.B. Green and J.H. Schwarz, Phys. Lett. **136B** (1984), 367

22. M.B. Green and C.M. Hull, QMC-89-9 (1989)
    R.E. Kallosh, Cern-TH-5355 (1989)
    U. Lindström, M. Rocek, W. Siegel, P. van Nieuwenhuizen, E.A. van de Ven and J. Gates, ITP-Stony Brook (1989)

23. I. Bengtsson, M. Cederwall and N. Linden, Phys. Lett. **203B** (1988), 96

24. T.F. Jordan and N. Mukunda, Phys. Rev. **132** (1963), 1842

25. L. Brink and J.H. Schwarz, Phys. Lett. **100B** (1981), 310

26. L. Brink, M. Henneaux and C. Teitelboim, Nucl. Phys. **B293** (1987), 505

27. A. Galperin, E. Ivanov, S. Kalitzin, V. Ogievetsky and E. Sokatchev, Class. Quant. Grav. **1** (1984), 469; **2** (1985), 155

28. E. Nissimov, S. Pacheva and S. Solomon, Nucl. Phys. **B297** (1988), 369

29. W. Siegel, Nucl. Phys. **B263** (1985), 93

30. I. Bengtsson, Phys. Rev. **D39** (1989), 1158

31. A.K.H. Bengtsson, I. Bengtsson, M. Cederwall and N. Linden, Phys. Rev. **D36** (1987), 1766
    I. Bengtsson and M. Cederwall, Nucl. Phys. **B302** (1988), 81

32. T. Shirafuji, Progr. Theor. Phys. **70** (1983), 18

33. M. Cederwall, ITP-Göteborg **89-15** (1989)

CHIRAL SYMMETRY AND CONFINEMENT

T. Goldman

Theoretical Division
Los Alamos National Laboratory
Los Alamos, NM 87545

Two principle features underlie the appearance of (approximate) chiral symmetry in hadronic systems. The first is that the conventional Dirac bispinor description of massless quarks hides the fact that this object is a direct sum of inequivalent representations of the Lorentz group. In a standard notation:

$$R[\psi] = (\tfrac{1}{2}, 0) + (0, \tfrac{1}{2}) \tag{1}$$

In general, there is an unrestricted phase between the two component representations which must be assigned physical meaning, if possible. For Majorana fermions, it is self-conjugacy (an operation outside the connected part of the Lorentz group) which fixes this meaning leaving only two independent degrees of freedom. For massive, charged fermions, the constraint arises (actually similarly) from the equality of the mass of a particle state of fixed spin (up, say) and the same spin antiparticle state.

In the charged, but massless, case, however, this last constraint vanishes, due (in one way of describing it,) to the absence of a rest frame in which to implement it. The phase loses physical meaning, and this fact is reflected in the chiral phase freedom (invariance) of the two (now unrelated within the proper orthochronous Lorentz group) components. Spontaneous chiral symmetry breaking (χSB) destroys this invariance by coupling the two representations by a scalar field vacuum expectation value (vev).

The second essential feature is that the dynamics must be chiral invariant as well. QCD satisfies this requirement by coupling equally to left and right chiral projections of the (nominally vector) color current (V$\pm$A) and so, to the separated components of the Dirac bispinor. (Conversely, the weak interactions explicitly violate chiral symmetry even without the formation of quark masses via the vev of the Higgs' scalar.] Chiral symmetry breaking, however, is now guaranteed: The attraction in the color singlet, Lorentz scalar channel between massless quark and antiquark must reduce the invariant mass-squared of the lowest state below the threshold value of zero. The resulting tachyonic values describe an unstable vacuum at zero vev, a conundrum cured by developing a quark vacuum condensate. The resulting effective Lagrangian of quark composite meson states is written as[1]

$$\begin{aligned}
L_q = &+ m^2 \left\{ (\sigma^2 + (\eta)^2 + \sum_j (\alpha_j)^2 + \sum_j (\pi_j)^2 \right\} \\
&- \lambda_1 \left\{ (\sigma^2 + (\eta)^2 + \sum_j (\alpha_j)^2 + \sum_j (\pi_j)^2 \right\}^2 \\
&- \lambda_2 \sum_j \left\{ (\sigma \alpha_j + \eta \pi_j) - (f_{jk\ell} \alpha_k \pi_\ell) \right\}^2 + \text{kinetic} + \ldots \tag{2}
\end{aligned}$$

where terms with higher powers of momenta and fields have not been written out explicitly, and f_{ijk} are the structure constants of the flavor group.

When the σ field is shifted by the vev,

$$\langle\sigma\rangle = 2\mu^2/\lambda_1 \tag{3a}$$

$$\tilde{\sigma} = \sigma - \langle\sigma\rangle \tag{3b}$$

L_q can be rewritten as

$$L_q = \ldots + 0(\sum_j \pi_j^2 + \eta^2) + \ldots + \text{kinetic} + \ldots \tag{3c}$$

Where we have explicitly noted the masslessness of the pseudoscalar (composite) bosons which has developed in accordance with the Goldstone theorem. In addition, instantons^2 contribute a term of the form

$$L_I = +\lambda_3 \, [\sigma^2 + \sum_j \pi_j^2 - \eta^2 - \sum_j \alpha_j^2] \tag{4}$$

which breaks the $U(1)_{axial}$ symmetry. The effects of the term include a shift in $\langle\sigma\rangle$, and a non-zero mass for the η, but the π_j remain massless.

We wish to point out here that a strong parallel exists in the sector of physical gluonic states, despite a markedly different initial appearance. [We ignore confinement, for the moment, just as was done above for quarks.] Although gluons appear in the (1/2,1/2) representation, the physical color electric (E_a) and magnetic (B_a) fields in the covariant stress tensor ($G_a^{\mu\nu}$) and its dual ($\tilde{G}_a^{\mu\nu}$) actually fill out a representation

$$R[G] = (1,0) \oplus (0,1) \tag{5}$$

with the same structure as $R[\psi]$ in Eq. (1).

The phase freedom between the two three-component representations has been implicitly recognized by the separation of solutions of the (covariant, source free) field equations into self-dual and anti-self-dual components. Thus, unless one refers to the underlying (1/2,1/2) representation, there is no Majorana-like constraint that may be imposed on Eq. (5). As in electrodynamics, the needed self-conjugacy can only be imposed by ruling out (independent) point magnetic (or electric) sources, with their opposing properties under parity. [Nor, obviously, is there any mass term to produce a Dirac-like constraint.]

Since the same dynamics is involved, it is immediate that pairs of (physical) gluons will bind to form tachyonic states, filling representations of a chiral symmetry entirely analogous to that for quarks, and differing only due to their being based on physically distinct degrees of freedom. In this case, however, the (color) octet channel is attractive as well as is the color singlet. [Any mixing with the repulsive (flavor singlet) color octet quark-antiquark channel will only serve to lower the lowest state, and so will not qualitatively affect the following argument.] Thus, (tachyonic) chiral $U(1) \times SU(3)$ color multiplets of gluonic mesons will be formed. The overall symmetry can not be a chiral $U(3)$ since the attraction in the color octet channel, while still stronger than for color singlet quarks and antiquarks, is weaker than for the color singlet combination of gluons.

As for quarks, this binding of massless objects must resolve its tachyonic tendencies by development of an SU(3) color singlet vev. Paralleling the quark notation, the effective Lagrangian of the (gluonic) composite meson states can be written as

$$L_G = + \mu^2 \left\{ |S_o + iP_o|^2 + \varsigma^2 \sum_{a=1}^{8} |S_a + iP_a|^2 \right\}$$

$$- \chi_1 \left[|S_o + iP_o|^2 + \varsigma^2 \sum_{a=1}^{8} |S_a + iP_a|^2 \right]^2$$

$$- \chi_2 \sum_a \left\{ (S_o S_a + P_o P_a) - \varsigma^2 f_{abc} S_b P_c \right\}^2$$

$$+ \text{kinetic terms} + \ldots \qquad (6)$$

where ς reflects the effect of the difference in U(1) and SU(3) strengths (and possible mixings with $q\bar{q}$ color octet components) referred to above.

As for the quark composite states, when S_o is shifted by the vev

$$<S_o> = g \qquad (7a)$$

$$\tilde{S}_o = S_o - <S_o> \qquad (7b)$$

L_G can be rewritten as

$$L_G = \ldots + O(P_o^2 + \sum_a P_a^2) + \text{kinetic} + \ldots \qquad (7c)$$

Note here that despite the separation of the U(1) factor, the pseudoscalar bosons all remain massless.

An addition corresponding to the instanton term now requires integrating over stationary (classical) quark configurations. These are not known to exist. If they did, however, their effects would be similar to that above: the value of $<S_o>$ would be shifted, and P_o would become massive.

These two sectors (2) and (6) would remain isolated, for massless quarks, due to the properties of perturbative QCD. However, non-perturbative effects, such as those produced by instantons do couple the two chiral sectors. That is, while a massless quark may not perturbatively annihilate with its antiquark in a J=0 state to produce two gluons, instantons couple a quark from $(\frac{1}{2},0)$ to an antiquark from $(0,\frac{1}{2})$. Thus, gluonic fluctuations coupled to the instanton are, in turn, coupled to a massless quark-antiquark pair (in a J=0 state among others). This allows spontaneous χSB from each sector to infect the other. To wit, we must add a term like

$$L_M = -\psi \left[|\sigma + i\eta|^2 + \sum_j |\alpha_j + i\pi_j|^2 \right] \left[|S_o + iP_o|^2 + \sum_a |S_a + iP_a|^2 \right] + \ldots \qquad (8)$$

It is now convenient to determine the vev's and mass terms by following the mixing of σ and S_o. Rather than carry the full complexity, we will keep only the singlet scalars and the (massless) multiplet of pseudoscalar states. In this minimal form, we consider

$$V_T = \lambda(\sigma^2 + \Sigma_j \pi_j^2 - v^2)^2 + \chi(S_o^2 + \Sigma_a P_a^2 - g^2)^2$$

$$+ \psi(\sigma^2 + \Sigma_j \pi_j^2 - w^2)^2 (S_o^2 + \Sigma_a P_a^2 - h^2) \tag{9},$$

the total effective potential in the σ, π_j, S_o, P_a field space. If the quark sector were isolated, $v = \langle\sigma\rangle$ would occur and similarly $g = \langle S_o\rangle$ for an isolated gluon sector. In the mixing term, we allow $w \neq v$, $h \neq g$ for full generality, and $h=w=0$ could also occur. Note that the λ and χ must be positive, although the sign of ψ is not determined in (9).

Our argument is that σ and S_o, since they have the same quantum numbers, and mix as argued above, form a system like the two real components of a complex Higgs' scalar: The effective potential in σ, S_o space forms a trough surrounding the origin, and only a specific combination acquires a vev. Let $c = \cos\theta$, $s = \sin\theta$,

$$\sigma = cz + sy \tag{10a}$$

$$S_o = -sz + cy \tag{10b}$$

and choose

$$\langle z \rangle = \Lambda \; ; \; \langle y \rangle = 0 \tag{11}$$

(by definition). The combination is messy, but some algebra shows that $\partial V_T/\partial z = 0$ for $y=0$ when $z=0$ or when

$$\langle z^2 \rangle = \frac{2\lambda c^2 v^2 + 2\chi s^2 g^2 + \psi(s^2 w^2 + c^2 h^2)}{2(\lambda c^4 + \chi s^4 + \psi s^2 c^2)} \tag{12},$$

which fixes $\Lambda^2 = \langle z \rangle^2 = \langle z^2 \rangle$. Of course, we must also have $\partial V_T/\partial y = 0$ at the same point. This fixes the angle θ via

$$2\lambda(c^2\Lambda^2 - v^2) + 2\chi(g^2 - s^2\Lambda^2) + \psi(w^2 - h^2 + [s^2 - c^2]\Lambda^2) = 0 \tag{13}$$

It is then straightforward to show that

$$m_y^2 = 2\lambda s^2(3c^2\Lambda^2 - v^2) + 2\chi c^2(3s^2\Lambda^2 - g^2)$$

$$+ \psi\{[(c^2 - s^2)^2 - 2c^2s^2]\Lambda^2 - (c^2 w^2 + s^2 h^2)\} \tag{14}$$

Note that $m_{\pi_j}^2 = 0 = m_{P_a}^2$ in the general case. This occurs because the mixing is entirely in the U(1) sectors.

Returning to the σ and S_o, we recognize that both acquired mass in the usual way, but that the ψ-term induced off-diagonal terms in the mass matrix for the two fields. The case $w=v$, $h=g$ is a little easier to follow, and the masses of the eigenstates are then:

$$m_z^2 = \frac{8(\lambda v^4 + \psi g^2 v^2 + \chi g^4)}{v^2 + g^2} \qquad (15a)$$

$$m_y^2 = \frac{8(\lambda - \psi + \chi) g^2 v^2}{v^2 + g^2} \qquad (15b)$$

Whether or not the lower mass state is composed mostly of glue or of quarks now depends on the detailed parameter values. It is apparent, however, that a 50%-50% mix is quite reasonable. We note in passing that this has significant consequences for the interpretation of the value of the σ-term of the purely quark model analysis.

We conclude with some comments regarding the P_a-states. Their masslessness is, of course, not a problem, since they are presumably confined by the usual color confinement mechanism. In this regard, they almost solve the confinement problem. For if it were only the S_a-scalars which were massless, the t-channel exchange of such a color octet between s-channel (say [heavy] quark and antiquark) sources of color would indeed provide a scalar potential. Since the effective coupling to the sources is $O(\alpha_s^2)$, it also grows at small (t-channel four-momentum-transfer-squared) q^2. Thus, similar conditions would apply to those Ball and Zachariasen suggest for t-channel gluon exchanges: a massless exchanged octet with diverging coupling. In this way, a linear scalar potential which confines color (including color octet and higher representation sources, not just quarks) could be obtained, with a color vertex structure parallel to that of single gluon exchange.

Unfortunately, it is the pseudoscalars which are massless, the Goldstone bosons of a broken <u>chiral</u> symmetry. It has never been suggested that the confining potential is a pseudoscalar! A vector O(9) symmetry of the states would be preferable, as the S_o (or mixture) vev would break it to O(8) and provide us with the desired (octet of) massless scalars. If nothing else, the $SU(3)_c$ invariance of the determinant of the octet of scalar and pseudoscalar glueball states explicitly violates such a symmetry. Thus, despite coming tantalizingly close, we are apparently still not able to link confinement to χSB in a transparent manner.

The author wished to thank A. W. Thomas, R. Crewther, K. R. Maltman, and G. J. Stephenson, Jr., for helpful conversations. This work was supported in part by the U. S. Department of Energy and the University of Adelaide.

REFERENCES

1. M. Gell-Mann and M. Levy, Nuovo Cimento <u>16</u> (1960) 53; B. W. Lee, Nucl. Phys. <u>B9</u> (1969) 649.

2. G. 'tHooft, Phys. Rev. D<u>14</u> (1976) 3432; Phys. Reports <u>142</u>, (1986) 357.

3. J. S. Ball and F. Zachariasen, CalTech reports (unpublished); M. Baker, J. S. Ball, P. Lucht, and F. Zachariasen, Phys. Lett. <u>89B</u> (1980) 211.

THE ORIGINAL FIFTH INTERACTION

Yuval Ne'eman [*] [#]

Raymond and Beverley Sackler Faculty of Exact Sciences

Tel-Aviv University, Tel-Aviv, Israel

and

Center for Particle Theory, University of Texas, Austin[**]

This essay is dedicated to Murray Gell-Mann upon his sixtieth birthday. I assume he will be particularly gratified to read that the inspiration for a new look at the 1964 work came during a visit to the College of Judea and Samaria at Ariel and to the "Nir" Yeshiva in Hebron. Also, while travelling in the area, I came across one of the five types of pheasant-like birds (Hebrew "HoGLaH") that he had enumerated to me on our 1967 tour of the Negev. There were no Malayalam-speaking immigrant Cochin Jews around this time as there had been at Nabatean Mamshit (Roman "Mempsis", and in Arabic "Kurnub") but I did visit Tat-speaking Daghestani "mountain Jews" in the Dothan valley in northern Shomron (Samaria). It is interesting that they have become important producers of goose-liver which is exported to France. Aside from Tat proper, they use "Judeo-Tat", a Tat-based "Yiddish".

---

[*]Wolfson Chair Extraordinary in Theoretical Physics
[#]Supported in part by the Binational USA-Israel Science Foundation, grant 87-00009/1.
[**]Supported in part by USDOE grant DE-FG05-85ER40200

QUANTUM FIELD THEORY, STRONG INTERACTIONS AND SU(3) IN 1964

In 1963-65, I was at Caltech as Murray's guest. We had met at CERN during the 1962 conference, at the end of the rapporteur session on strange particles. I have related elsewhere the events of that day [1] and Gerson Goldhaber has added a witness-participant account [2]. Anyhow, the outcome was a two-year stay at Caltech and the beginning of a lasting friendship.

At Caltech, Gell-Mann was in the last stages of a series of studies, mostly with Marvin Goldberger, Francis Low and Fred Zachariasen, but also with Egon Marx and V. Singh [3], aiming at a reconciliation between Relativistic Quantum Field Theory (RQFT) and "S - matrix Theory" (SMT).

RQFT had been so successful in describing Quantum Electrodynamics (QED) that it seemed worth trying to preserve it as the overall description of the Physics of Particles and Fields (PPF), even though it was encountering great difficulties in its application to the other interactions. In the description of the Weak Interaction (WI), the "current x current" Lagrangian as identified by Sudarshan and Marshak - and by Feynman and Gell-Mann - had a dimensionality D = - 6, with the Fermi constant G_F supplying D = 2, to reproduce a dimensionless action after the d^4x integration. A coupling with D = 2 meant that in a perturbative treatment, one would require an infinite number of counter terms with different dimensionalities - something certainly beyond conventional renormalization methods; indeed, the WI renormalization was only accomplished in 1971, when 't Hooft proved [4] that the Weinberg-Salam [5] Lagrangian is renormalizable. That theory, aside from providing a description of the precise way in which electromagnetic and "weak" interactions mix (the term "unification" really does not fit this case), also resolved that dimensionality problem. D = 2 is the dimensionality of an area, but it is also that of the inverse of a squared mass. The Fermi constant's dimensionality turned out to be due to the presence of the mediator-boson (the W) propagator; a process such as beta-decay, described by the Fermi constant in first order, now turned out to represent a second order process in the new theory, with contracted W meson lines resulting in a propagator, with the usual squared mass in the denominator:

$$G_F = (e^2\sqrt{2})/8M_W^2 sin^2\theta$$

e is the (dimensionless) charge of the electron, θ is the Weinberg angle. Let me remark that in a model in which I have tried [6] to "really unify" the Weinberg-Salam theory through the embedding of SU(2) x U(1) in the supergroup SU(2/1), one has $sin^2\theta = 1/4$ and the mass of the Higgs meson [7] is $M_H = 2M_W$.

For the Strong Interaction (SI), the baryon-meson pseudoscalar Yukawa Lagrangian had a dimensionless coupling, but its value $g \approx 15$ made it impossible to use the perturbation series here too. Both problems appeared to make RQFT useless, but not wrong in principle. However, Geoff Chew had taken the next step and declared it plain wrong. It had been shown that in various RQFT, Unitarity appeared to be lost off-mass-shell, and that - if unremedied - meant RQFT was wrong, its success in QED notwithstanding. Chew was prepared - in view of that success - to give RQFT another chance in the Weak Interaction, but certainly not in the Strong.

Gell-Mann and his collaborators were trying to show that the main novel feature displayed by SI, i.e. the Regge "trajectories" in the (analytically- continued) complex angular momentum plane, could be reproduced in particular types of RQFT. The Regge trajectories were also described by straight lines in the Chew-Frautschi diagram, plotting the real part of angular momentum against the square of the center of mass energy (in the direct channel) on the positive side, or the square of the momentum transfer (in the crossed exchange channel), for negative values of that variable "t". At integer or half-integer values of the angular momentum real part, this would correspond in the direct channel to the squared mass, for sequences of resonance particles lying along these lines.

The cross-channel $E \leq 0$ part dominates the scattering.

All these features could be reproduced in some RQFT. Gell-Mann and his collaborators were studying the $Z_3 \to 0$ limit (vanishing of the coupling renormalization constant) in these theories. It is gratifying to remember that this approach was indeed vindicated 10 years later, after 't Hooft had managed to complete [4] the renormalization of the Yang-Mills interaction (YM) [8]; here this corresponds to Asymptotic Freedom (AF), discovered by 't Hooft, by Gross and Wilzcek and by Politzer [9]. The reestablishment of the consistency between RQFT and SMT required a picture in which the hadrons are composite, so as (1) to explain the Regge sequences in analogy to atomic or nuclear-like excitations, and (2) to fit the "bootstrap" approach, in which hadrons are "made of" each other, by reinterpreting high-energy scattering as "rearrangements" of the constituent quarks (a picture already seen in the Harari-Rosner [10] Duality Diagrams, further detailing the Finite Energy Sum Rules [11]).

All of this was happening in the aftermath of our (simultaneous and independent) discovery of SU(3) ("Unitary Symmetry"). As a matter of fact, both the supporters of the SMT and some of the RQFT faithful were spreading heretical (or at least agnostic) arguments against SU(3), in the SI context. Basically, they were objecting to the successes of predictions relating to the symmetry breaking, derived from a first-order formalism, such as the Gell-Mann/Okubo mass formula, including the prediction of the mass of the Omega-Minus, as discussed, for instance, in the Yang-Oakes article [12]. In the latter case, eleven answers were published within three months, showing how in that case, the algebraic features could survive. And yet, it was difficult to see why in a strong coupling theory, the SU(3)-breaking terms should not be as strong in the "n"th order as in the first, and why should formulae based on simple-minded first-order perturbation theory work so well.

Gell-Mann's invention [13] of Current Algebra (CA) partly resolved the general problem of the "visibility" of a SI symmetry. The algebraic generators of the SI symmetry, the global SU(3) (or SU(3) x SU(3)) charges had as charge-current densities local operators corresponding to the currents of the "weak" or electromagnetic interactions. The current commutators could be preserved, in the spirit of Matrix Mechanics; their definition involved the SI to all orders, but the WI, QED and Gravity to first order only, and this was "Kosher" (American Ashkenazy-Hebrew pronounciation; "Casher", pronounced "Kah-sher" in the Sepharadi dialect selected by E. Ben-Yehuda for modern "Israeli" Hebrew). CA in many cases constituted an effective alternative to RQFT. And yet, even after the introduction of CA, it was not clear why SU(3) should only be "slightly" broken in a hadron scattering experiment, or in the mass formula: with the Hamiltonian density defined by the extremely weak gravitational coupling in first order, one could understand the Coleman-Glashow sum rules for electromagnetic mass differences or magnetic moments, applying to that Hamiltonian density a first order electromagnetic effect; but there was no such interpretation for the Gell-Mann/Okubo formula, as long as there was no similar derivation of the SU(3)-breaking effect as resulting from a "weaker" force, also considered to first order.

THE FIFTH INTERACTION (1964)

Pondering these issues, I realized that one possible way of explaining how SI could be strong, and yet provide us with a well-ordered and perturbative symmetry-breaking, might consist in separating out *dynamically* the two SU(3)-kinematically different "components" in the SI.

One component, the SU(3)-invariant piece, would represent the "truely strong" SI, with couplings bounded only by Unitarity, with Regge excitations and Analyticity, obeying bootstrap mechanisms and justifying the use of SMT.

The other component, the SU(3) breaking interaction would, in such a picture, represent a completely different force, describable by *RQFT* and a perturbative treatment. Moreover, considering the similarity in magnitude between the value of the *SU(3)* breaking term in the hadron mass Lagrangian (the mass of the "s" quark), and that of the muon mass, I suggested that the new force might be responsible for both.

I published these ideas [14]. We also included that article in our book "The Eightfold Way" [15]. I called the new force (and the article in the Physical Review) "The Fifth Interaction", for obvious reasons. I was therefore surprised when twenty two years later [16] the world press (including the Israeli dailies) carried a story about a "Fifth Force", that was said to have been discovered by a group at Purdue and Seattle. I called Murray from Tel-Aviv, and Helen got him out of his class in the middle of his Wednesday lecture. He was surprised, but explained to me that this was a supposed medium-range correction to Gravity, arising out of a new analysis of the 1922 Eötvös experiment. This could still be connected with hypercharge (i.e. B+S, where "B" is baryon number and "S" is Strangeness), but it did not have to. Apparently, the discoverers (the evidence for their effect does not appear to be very strong now) either did not know of my 1964 paper and naming, or they might have decided that my idea had been wrong anyhow, so that the name was again available for appropriation.

The present essay purports to show that the idea of precisely that "Fifth Interaction" of 1964 vintage is still very much alive - in fact that its first postulate has been proven to be right and is now complementary to the Standard Model.

GENERATIONS, QCD AND THE FIFTH

With the postulation of QCD [18,9,10], the above separation was indeed adopted. The $SU(3)$ invariant component of the SI is supposed to be due to a Yang-Mills [8] force coupled to $SU(3)_{colour}$, a group that commutes with the SU(3) of Unitary Symmetry. Analyticity, duality, unitarity-bound couplings - all of this is assumed to result from

QCD.

The Standard Model, however, cannot explain the masses of the quark and lepton fields. These are considered as input parameters. There is also no reason for the existence of more than one "generation" (or "family"). In the table of "fundamental" fields of matter,

$N = 0$ $\qquad\qquad\qquad N = 1 \qquad\qquad\qquad N = 2$

Quarks

$(1,0)\ U = 4/3\ I = 0$
$u_R^{2/3} \rightarrow \qquad\qquad c_R^{2/3} \rightarrow \qquad\qquad t_R^{2/3} \rightarrow$
$u_B^{2/3} \rightarrow \qquad\qquad c_B^{2/3} \rightarrow \qquad\qquad t_B^{2/3} \rightarrow$
$u_Y^{2/3} \rightarrow \qquad\qquad c_Y^{2/3} \rightarrow \qquad\qquad t_Y^{2/3} \rightarrow$

$(1,0)\ U = 1/3\ I = 1/2\ I_z = +1/2$
$u_R^{2/3} \leftarrow \qquad\qquad c_R^{2/3} \leftarrow \qquad\qquad t_R^{2/3} \leftarrow$
$u_B^{2/3} \leftarrow \qquad\qquad c_B^{2/3} \leftarrow \qquad\qquad t_B^{2/3} \leftarrow$
$u_Y^{2/3} \leftarrow \qquad\qquad c_Y^{2/3} \leftarrow \qquad\qquad t_Y^{2/3} \leftarrow$

$(1,0)\ U = 1/3\ I = 1/2\ I_z = -1/2$
$d_R^{-1/3} \leftarrow \qquad\qquad s_R^{-1/3} \leftarrow \qquad\qquad b_R^{-1/3} \leftarrow$
$d_B^{-1/3} \leftarrow \qquad\qquad s_B^{-1/3} \leftarrow \qquad\qquad b_B^{-1/3} \leftarrow$
$d_Y^{-1/3} \leftarrow \qquad\qquad s_Y^{-1/3} \leftarrow \qquad\qquad b_Y^{-1/3} \leftarrow$

$(1,0)\ U = -2/3\ I = 0$
$d_R^{-1/3} \rightarrow \qquad\qquad s_R^{-1/3} \rightarrow \qquad\qquad b_R^{-1/3} \rightarrow$
$d_B^{-1/3} \rightarrow \qquad\qquad s_B^{-1/3} \rightarrow \qquad\qquad b_B^{-1/3} \rightarrow$
$d_Y^{-1/3} \rightarrow \qquad\qquad s_Y^{-1/3} \rightarrow \qquad\qquad b_Y^{-1/3} \rightarrow$

Leptons

$U = -1\ I = 1/2\ I_z = +1/2$
$\nu_e^0 \leftarrow \qquad\qquad \nu_\mu^0 \leftarrow \qquad\qquad \nu_\tau^0 \leftarrow$

$U = -1\ I = 1/2\ I_z = -1/2$
$e^- \leftarrow \qquad\qquad \mu^- \leftarrow \qquad\qquad \tau^- \leftarrow$

$U = -2\ I = 0$
$e^- \rightarrow \qquad\qquad \mu^- \rightarrow \qquad\qquad \tau^- \rightarrow$

N is "Seriality", the generation number, which is outside of SU(3) x SU(2) x U(1). We have used N = 0 for the "ground state", in the assumption that the higher levels represent the inclusion of a new type of charge, absent in that ground level. The Standard Model quantum numbers are for $SU(3)_{(c)} : (\lambda, \mu)$ for the representation, and $y_{(c)}, i_{(c)}, i_{z(c)}$ for the state; the latter are however replaced by the indices R,B,Y denoting the "colours", i.e. (1/3, 1/2, +1/2), (1/3, 1/2, -1/2) and (-2/3, 0, 0) respectively.

In addition, we have U, the "weak hypercharge", I the "weak isospin", I_z its third component, representing the U(1) x SU(2) quantum numbers (in fact a U(2)); $\leftarrow$ and $\rightarrow$ denote left- and right-chiral states. Note that d and s represent the Cabibbo-mixed linear combinations.

It is not obvious that the Kobayashi-Maskawa [18] inter-generation mixing is also due to the Fifth Interaction, although the usual assumption is that it reflects a limited degree of non-commutativity between the N-exciting Fifth and QCD. However, it could also correspond to yet another interaction, and in this study we make that assumption. Note that in 1967, Cabibbo [19] initiated an attempt to derive the 15 deg. angle from a "spontaneous" breakdown of SU(3): writing an SU(3)-invariant equation for the meta-theory, inserting a small SU(3)-breaking in the λ_8 direction and hoping to get as a result the Cabibbo angle, as a rediagonalization of the Hamiltonian. The method was developed by Brout [20] and by Michel and Radicati [21], who generalized it to any Lie group. The result was consistently negative, i.e. the 8th component-directed breaking of SU(3) might have caused $\theta = 0$, or $\pi/2$, but not $\pi/8$. This is why we prefer to assume that the Kobayashi-Maskawa mixing matrix is not the result of the Fifth.

In the discussion of "beyond-SM physics", as conducted in the Seventies and Eighties, three types of effects are described as "Higgs-type interactions": the quark and lepton masses (u.v. masses for the quarks), the Kobayashi-Maskawa generalized Cabibbo mixings and the non-vanishing vacuum-expectation values of the Higgs field (or fields). In this "Fifth Interaction" version, following the argument quoted in the previous paragraph, only the first set is considered as arising from that force. The N-mixing that occurs in the low-energy "hadron-constituent" quarks will thus be assumed to be superimposed "later".

SERIALITY CONSERVATION AND DYNAMICAL CHARACTERISTICS OF THE FIFTH

We note the specific - and unusual - features of the N-excitations:

(a) There is one very important conclusion from this picture of N as a precise quantum number. *The Fifth Interaction is then the origin of the conservation of charm, of strangeness, of "truth" and of "beauty", and also of the muonic and tauonic lepton numbers. In fact, these six conservation laws now become one single law, that of N-conservation.*

(b) There is one other issue that has to be related to the Fifth Interaction: *the compositeness of quark and leptons.* This is very plausible, since the N levels have to represent a band of excited states involving the same set of constituents for each row of the table. The fact that the entire SM column is "evenly" N-excited seems to indicate that the relevant constituents are the "preons" "rishons" or "haplons" that make up the entire set of 15 quark and lepton fields in a generation. This indication presumably reflects the nature of the interaction that glues together three rishons, for instance in the Harari-Shupe model [22], to make a quark or a lepton. Should this force be of the Yang-Mills type, a confinement as in QCD would imply a simple group, rather than the Abelian force that we considered in [14].

As a matter of fact, three possibilities should be analyzed:

(1) Abelian $U(1)$,

(2) $SU(2)$, with generators raising and lowering N by one unit at a time, and

(3) SU(3), with the possibility of raising N by two units in one application of the generators.

In this last version, one could define an SU(6) from the envelope of the tensor product of this $SU(3)_{fifth}$ and the "weak" SU(2). Thinking of the analogy with the Gursey-Radicati SU(6), the $SU(3)_{fifth}$ plays the role of "our" original SU(3), and the $SU(2)_{weak}$ plays the role of the spin, in the Gursey-Radicati SU(6). The KB mixings would then be equivalent to a redefinition of the spin within SU(6)..

Note that with no indications with respect to direct first-to-third generation transitions, and with the requirement of confining capabilities, we would rather opt for (2), i.e. an SU(2).

Note also that SU(3) would be appropriate in a model such as that of Harari and Shupe [22], in which the "hyper-colour" displays saturation for 3-constituent compounds, as in the case of "colour"-SU(3).

Any treatment of these constituents and their glueing interaction will have to obey 't Hooft's constraints relating to the anomalies [23].

(c) The spin is not excited, within the presently observed energy regime. The first and second vibrational levels appear before the first rotational excitation, in the language of atomic or molecular physics (where the ordering is inverted, the rotational energy increment being 1/50 th of the vibrational one). Alternatively, we can think of radial modes as in the hadron spectrum. The nucleon, indeed, has a first "recurrence" with spin $J = 1/2^+$ (above the ground state N(940)) at N(1440), and then a second one at N(1710) etc.., while its first rotational excitations, with $J = 3/2^-$ and $J = 5/2^+$ are respectively at N(1520) and N(1680). This is closer to what we are after here. However, QCD does excite the spins as well, forming Regge sequences. At the quark-lepton level, they must appear at much higher energies. In any case, we have to look for a mechanism favouring "radial" excitations over the rotational.

(d) The excitations appear to be extremely stable, as exemplified by the bounds on $\mu \not\to e + \gamma$.

(e) We have at this stage no observed example of N- raising or N- lowering processes, even when the conservation of Seriality is not violated, as in the process,

$$\Lambda^0(\text{or } s^{-1/3}) \to n^0(\text{or } d^{-1/3}) + e^- + \mu^+ \qquad (1)$$

or also

$$\Lambda^0(\text{or } s^{-1/3}) \to n^0(\text{or } d^{-1/3}) + e^+ + \mu^- \qquad (2)$$

Note, however, that in our approach, only the second process is allowed, since $s^{-1/3}$ has $N = 1$, and so does μ^-, whereas μ^+ has $N = -1$. Similarly, we would expect

$$\Sigma^+(\text{or } s^{-1/3}) \to p(\text{or } d^{-1/3}) + e^+ + \mu^- \qquad (3)$$

Presumably, both (d) and (e) could be due to the mediating boson X having a large mass. Since we do not have to cope, in this ("horizontal") Fifth Interaction, with bounds such as those imposed by the proton lifetime on the ("vertical") GUTs, it seems that even masses of the order of $10^5 GeV$ are not yet excluded. There are numerous processes of the above type, and it would be useful to determine the present upper bounds on the cross-sections, which would also determine the lower bound on the mass of the X boson.

(f) The mediating bosons, whether N- raising or N- lowering, are electrically neutral. However, they presumably do not mix with the electromagnetic field because of N- conservation, as they carry N- charge. The photon in $\mu \to e + \gamma$ does not mix with X, so that this process should stay forbidden even when the above $s \to \mu$ transition is allowed.

(g) Many authors have commented on the general features of the mass spectrum of the above table; see for example Fritsch's article in these proceedings.

REFERENCES

1. Y. Ne'eman, in "Symmetries in Physics, 1600 - 1980", M.G. Doncel et al. edts., University of Barcelona and World Scientific Pub., Barcelona (1987) pp. 510-555.

2. G. Goldhaber, in "From SU(3) to Gravity", E. Gotsman and G. Tauber edts., Cambridge University Press (Cambridge 1985) p. 103.

3. M. Gell-Mann and M.L. Goldberger, Phys. Rev. Lett. 9 (1962) 275; same authors, F.E. Low and F. Zachariasen, Phys. Lett. 4 (1963) 265; same authors and E. Marx, Phys. Rev. 133 (1964) B145; same authors, except for V. Singh replacing E. Marx, Phys. Rev. 133 (1964) 162.

4. G. t'Hooft, Nucl. Phys. B35 (1971) 167.

5. S. Weinberg Phys. Rev. Lett. 19 (1967) 1264; A. Salam, in Elementary Particle Theory, N. Svartholm ed., Almquist and Wiksells, Stockholm (1968)

6. Y. Ne'eman, Physics Letters B81 (1979) 190; D.B. Fairlie, Physics Letters B82 (1979) 97.

7. Y. Ne'eman, Physics Letters B181 (1986) 308.

8. C.N. Yang and R.L. Mills, Phys. Rev. 96 (1954) 191.

9. G. 't Hooft, unpub. D.J. Gross and F. Wilzcek Phys. Rev. Lett. 30 (1973) 1343; H.D. Politzer Phys. Rev. Lett. 30 (1973) 1346.

10. H. Harari Phys. Rev. Lett. 22 (1969) 562; J. Rosner, Phys. Rev. Lett. 22 (1969) 689.

11. R. Dolen, D. Horn and C. Schmid Phys. Rev. Lett. 19 (1967) 402; A.A. Logunov, L.D. Soloviev and A. Tavkhelidze Physics Letters B24 (1967) 81; K. Igi and S. Matsuda Phys. Rev. Lett. 18 (1967) 625.

12. R.J. Oakes and C.N. Yang, Phys. Rev. Lett. 11 (1963) 174; see also e.g. R.J. Eden and J.R. Taylor, Phys. Rev. Lett. 11 (1963) 516.

13. M. Gell-Mann, Phys. Rev., 125 (1962) 1067; Physics, 1 (1964) 63.

14. Y. Ne'eman, Phys. Rev. 134 (1964) B1355.

15. M. Gell-Mann and Y. Ne'eman, "The Eightfold Way", W.A. Benjamin Pub., New York (1964).

16. E. Fischbach, D. Sudarsky, A. Szafer, C. Talmadge and S.H. Aronson, Phys. Rev. Lett. 56 (1986) 3.

17. Y. Nambu, in Preludes to Theoretical Physics, A. de Shalit, H. Feschbach and L. Van Hove edts., North Holland Pub. (Amsterdam, 1966).

18. M. Kobayashi and T. Maskawa, Prog. Theor. Phys. 49 (1973) 652.

19. N. Cabibbo, in Hadrons and their Interactions, A. Zichichi ed., Academic Press (NY 1968).

20. R. Brout, Nuovo Cim. A47 (1967) 932.

21. L. Michel, in Group Representations in Math. and Phys., Batelle Rencontre, Seattle 1969, Springer Ver. p. 136.

22. H. Harari Physics Letters 86B (1979) 83; M.A. Shupe, Physics Letters 86B (1979) 87. See also Y. Ne'eman, Physics Letters 82B (1979) 69 and H. Harari and N. Seiberg, Physics Letters 98B (1981) 269.

23. G. 't Hooft, in Recent Developments in Gauge Theories, G. 't Hooft ed., Plenum Press, New York (1980).

THE MASS HIERARCHY OF LEPTONS AND QUARKS
AS A NEW SYMMETRY[*]

Harald Fritzsch

Sektion Physik der Universität München

and

Max–Planck–Institut für Physik und Astrophysik
— Werner Heisenberg Institut für Physik —

München, Germany

[*] Supported in part by DFG–contract Fr. 412/7–2

In the summer of 1970 I attended as a graduate student from MPI Munich the Brandeis Summer School on Theoretical Physics at Brandeis University. Afterwards I drove in a car which I had to deliver eventually in Long Beach, California, throughout the United States. This trip was not only my first encounter with the magnificent sceneries of the United States. On a short stay at the Physics Center in Aspen, Colorado, I met in a discussion with colleagues on problems of broken scale invariance Murray Gell—Mann for the first time.

The year 1970 was an exciting one in particle physics. After several years of frustration and little progress in experimental studies, the observation of the scaling phenomena in inelastic electron—nucleus scattering at SLAC had started a new era in particle physics. I had the hunch, like numerous other theorists, that the "SLAC scaling" might have something to do with scale invariance in field theory, the topic of my Ph. D. — thesis, which had been given to me by Heinrich Mitter at the MPI in Munich. In 1970 Gell—Mann was working, partially together with Peter Carruthers, on the problem of scale invariance and its breaking in hadron physics, a topic, which at a first sight seemed unrelated to the "scaling phenomenon" seen at SLAC. I remember a number of conversations I had with Murray at the Aspen Physics Center, in which we talked about possible connections. In those days, the time of duality and the dual resonance models, it was unpopular to talk about hadrons, strong interactions and field theory at the same time. Nevertheless Murray and I started to look at the problem in terms of field theory, especially of the quark—gluon models.

In one of our first discussions, after I had told Murray about my work in Munich, he mentioned that he dislikes to work with students, but prefers to work with postdocs, and at the same time he asked me to join Caltech for a while to start some sort of collaboration. After that I did not dare to disclose the fact that I had not yet finished my Ph. D., but promised to come to Caltech for a short time after my planned stay at SLAC for six months starting in the fall of 1970.

Murray and I started to collaborate in January 1971. I shall never forget the day of my arrival in Pasadena. I had taken an early morning flight from San José airport to Burbank. As the plane approached the Los Angeles area, the captain informed us that an earthquake had hit Southern California minutes ago. An hour later I observed how the earthquake had transformed Gell—Manns Caltech office like all others at Caltech into an example of chaos. Especially all the pictures on the walls were askew. Murray took great care to preserve the chaotic state of his pictures until today and likes to tell his visitors about its origin, mentioning in addition that the day of the big earthquake was also the first day of our collaboration.

(Caltech insiders claim that one morning somebody new and uninformed of the cleaning personnel straightened out the pictures, but Murray, gifted with an extremely good memory, put them back into their after–quake position.)

It happened by accident that our collaboration began at a time when particle physics started to approach a new age, the age of what is now inappropriately called the "standard model". Five years later, at the beginning of 1976, it had become clear that this model, a combination of the electroweak theory and QCD, was extremely successful in describing all particle physics phenomena.

In retrospect it is obvious why Gell–Mann's name will be linked forever with the slow, but rather systematic evolution of the standard model since the early fifties. All the important steps in this evolution listed below were either initiated by him or are linked with important contributions of him:

a) The correct interpretation of the strange particle phenomenon.
b) Early ideas on the concept of parity violation.
c) The proposition of the universal (V–A)–theory of the weak interactions.
d) The SU(3) flavor symmetry of the hadrons and their interactions.
e) The current algebra of the observable current densities.
f) The gauge theory of the electroweak interactions.
g) The concept of quarks.
h) The interpretation of the "scaling phenomena" in terms of quarks and gluons.
i) The gauge theory of QCD.

Without Gell–Mann the evolution of particle physics in the second part of our century would have proceeded differently, at least more slowly, and today we would not have arrived at the position we are now.

Today particle physics has reached a peculiar stage in its development. For the first time we are in the possession of a theory, the "standard model", which is able to describe all phenomena in particle physics. All interactions in nature except gravity can be reduced to two types of basic interactions, the strong QCD force among the quarks, and the electroweak interactions among the leptons and quarks. Thus one of the main goals in physics, a reduction of the multitude of natural phenomena to a small number of basic forces among a small number of fundamental matter constituents, has been reached. It is not excluded that the end of the road in subnuclear physics has been reached. Whether this is really the case or not, depends on, whether new phenomena are discovered in the future, which cannot be described in the terms of the "standard model". To some extent this has already happened in the sense that we can speak of an success of the "standard model" only if at least 18 free parameters are fitted appropriately, i.e. taken from observation. Four of these numbers can be denoted as strength parameters (coupling constants), which describe the absolute strengths of the interactions, among them the famous finestructure constant α, introduced by Sommerfeld long ago. To calculate any of these numbers remains an important task for the future, which might be done only, once a successful unification of the "standard model interactions" and gravity is achieved.

However we cannot exclude the possibility that these strength parameters are numbers of an historical origin, i.e. fixed by yet unknown cosmological processes shortly after the "big bang". In this case they can never be calculated, but are essentially of accidental nature. Of course, this possibility does not exclude that there might exist relations among the coupling constants similar to the ones obtained in a grand unified theory like the SU(5) or SO(10) model.

The remaining 14 parameters are mass parameters, one describing the mass scale of the weak bosons, nine for the masses of the three charged leptons and the six quarks, as well four for the weak interaction mixing matrix of the quarks. I think that this spectrum of mass parameters is a signal that the yet unknown physics beyond the "standard model" has its own specific dynamics, and it all depends whether we are able to read this signal in the right way.

In the beginning of the 60ies Gell-Mann has shown, how symmetries, especially the approximate symmetries in the mass spectrum of hadrons, could be used to deduce step after step the essential ingredients of the strong interactions, the quarks and gluons, confined by the QCD forces. Probably it is too early to claim that a closer inspection of the lepton-quark mass spectrum could result now in deriving the essential features of the internal dynamics responsible for this mass spectrum. But a first step in this direction could be made soon, perhaps has been made already.

A closer inspection of the mass spectrum tells us:
a) The mass spectra of the three flavor channels (charged leptons, u-type quarks, d-type quarks are almost entirely dominated by the mass of the member of the third generation.
b) The relative importance of the second generation decreases as we proceed upwards in the charge ($\mu \rightarrow s \rightarrow c$). In the lepton case the muon contributes

about 5.6 % to the sum of the masses, while in the charge $-1/3$ – channel the s–quark contributes only 3.2 %, and in the charge $+2/3$ – channel the c–quark contributes only 1.3 % (for $m_t = 100$ GeV).

c) The relative importance of the masses of the members of the first generation is essentially zero.

d) The entire mass spectrum of the leptons and quarks is dominated fairly well by the t–quark alone. For example, in the case $m_t = 100$ GeV the t–quark alone contributes 95 % to the sum of all fermion masses. All other quarks, mostly the b–quark, contribute only 5 %.

The spectrum exhibits clearly a hierarchical pattern: The masses of a particular generation of leptons or quarks are small compared to the masses of the following generation, if there is any, and large compared to the previous one if there is any. Furthermore another hierarchical pattern emerges if we consider the weak interaction mixing parameters. The mixing matrix, if written in terms of quark mass eigenstates, is not far from the diagonal matrix (no mixing). The mixing angles are typically rather small; the Cabibbo angle being the largest of all, is about 13°.

What kind of symmetry could one discuss in view of the observed lepton–quark mass spectrum? The observed two different hierarchies suggest that we are very close to a limit, which I like to call the "rank 1" – limit, in which both the u–type and d–type mass matrix can be diagonalized at the same time and in which they both take the diagonal form (0, 0, 1), multiplied by m_t or m_b respectively. Thus the masses of the first two generations vanish (the mass matrix has rank one), likewise all mixing angles. Of course, it depends on yet unknown details of the mass generation mechanism whether such a limit can be achieved in a consistent way. We simply assume that this is the case. In this limit there exists a mass gap: The third generation is split from the massless first two generations. Obviously nature is not far away from this limit, and therefore one is invited to speculate about the dynamical origin of such a situation. A mass matrix proportional to the matrix

$$\begin{bmatrix} 0 & 0 & 0 \\ 0 & 0 & 0 \\ 0 & 0 & 1 \end{bmatrix}$$

can always be obtained from another matrix, namely the one in which all elements are equal:

$$\begin{bmatrix} 1 & 1 & 1 \\ 1 & 1 & 1 \\ 1 & 1 & 1 \end{bmatrix}$$

by a suitable unitary transformation. Matrices of this type, which might be called "democratic mass matrices" have been considered in the past by various people, in particular by Harari and Weyers, by C. Jarlskog, by Koide and myself.

More recently a mass matrix of the type given above was considered by Kaus and Meshkov, Nambu and myself, where it was used as a starting point to construct the full mass matrices of the quarks, including the weak interaction mixing terms. It was also emphasized that such a matrix plays an important rôle in other fields of physics, where mass gap phenomena are observed:

a) In the BCS theory of superconductivity the energy gap is related to a "democratic" matrix in the Hilbert space of the Cooper pairs.

b) The pairing force in nuclear physics which is introduced in order to explain large mass gaps in nuclear energy levels has the property that the associated Hamiltonian in the space of nucleon pairs has equal matrix elements, i.e. it has a structure of the type given above.

c) The mass pattern of the pseudoscalar mesons in QCD in the chiral limit $m_u = m_d = 0$. In this limit the π^o and the η are massless Goldstone bosons, while the η' acquires a mass due to the gluon anomaly.

Once we write the mass matrices of the leptons and quarks in their "democratic" form, it is obvious that there exists a symmetry, namely the symmetry S_3 of permutations among the three different flavors. This symmetry suggests that one should consider the eigenstates of the quarks and leptons in this basis as the fundamental dynamical entities. Let us denote them as (l_1, l_2, l_3) and (q_1, q_2, q_3) respectively.

The heaviest lepton and quark, i.e. the τ-lepton, the t and b quarks, would be coherent states of the type:

$$\tau = \frac{1}{\sqrt{3}} (l_1 + l_2 + l_3) \quad \text{etc.}$$

In view of the scarce information we have at present about the internal dynamics of the leptons and quarks we do not know, whether this description of the fermions in terms of coherent states is more than a specific mathematical representation. In a composite model, for example, the fermion states f_1, f_2, f_3 would be those states which are "pure" in a dynamical sense, e.g. they have simple unmixed wave functions.

We remind the reader that also in the case of superconductivity and of the nuclear pairing force the mass eigenstates are coherent superpositions of "physical" states which are described by simple wave functions (e.g. the Cooper pairs in superconductivity).

Within our approach we see a solution to a problem, which has plagued many models of the physics beyond the standard model, the problem of the near masslessness of the first and to some extent also of the second generation. In the coherent state basis this is easily understood. For example, the electron state $e = 1\sqrt{2}\,(l_1 - l_2)$ is nearly massless, since there is a nearly complete cancellation of the l_1- and l_2- mass terms, as a consequence of the rank one structure of the dominant lepton—quark mass term.

Although this is not the place to discuss the details of the mass generation for the first two generations, it is important to note that in various dynamical schemes, in particular in one based on a composite structure of the leptons and quarks, the introduction of these small masses leads to slight breakings of the flavor conservation, especially in reactions associated with large momentum transfers. In particular the reaction $e + p \to \tau + X$, a reaction, which might be found soon at HERA, is of interest. Moreover rare decays of the Z_0 – boson like $Z_0 \to \bar{t}c$ or $Z_0 \to \bar{\mu}\tau$ should be looked for. If the t–quark is heavier than 92 GeV, one should look for the decays $t \to Zc$.

In this talk I have described a number of ideas which one might consider after looking at the pattern of masses exhibited in the lepton—quark mass spectrum. I have emphasized the rôle of symmetries in the space of the generations of the quarks in providing relations between the various mass eigenvalues and the mixing angles. An approach to the flavor problem and to the hierarchical mass spectrum of the leptons and quarks, based on the introduction of coherent states, was discussed. It was argued that the mass generation for the third lepton—quark generation is nothing but a gap phenomenon and is rather similar to the mass generation for the pseudoscalar mesons in QCD. Thus the third lepton—quark generation is somewhat distinct from the other ones. The same mechanism which leads to the mass generation causes the appearance of flavor changing effects; only in the absence of the lepton and quark masses of the first and second generation the various quark and lepton flavors are conserved.

If our interpretation of the mass gap seen in the lepton–quark spectrum is correct, it would mean that all mass gap phenomena seen in physics — superconductivity, nuclear pairing forces, QCD mass gap, lepton–quark mass spectrum — are due to an analogous underlying dynamical mechanism. The exploration of further details of this mechanism could lead soon to a deeper understanding of the physics beyond the standard model.

The present situation in particle physics is in my view somewhat similar to the one around 1960. At that time new symmetries were found among the particles, and it took many steps like current algebra, the SU(6) framework, the scaling behaviour in the deep inelastic region of lepton — hadron scattering and its interpretation in terms of the light–cone behaviour of current products etc. in order to arrive at the QCD–framework of our present time. Gell–Mann played the leading rôle in this development. His method, an ingenious combination of a deep theoretical insight and an intuitive flair for the phenomena to look for, is needed today in order to overcome the problems high–energy physics is facing at the present time.

Acknowledgement:

It is a pleasure to thank John Schwarz and Fred Zachariasen for organizing this memorable conference at Caltech, in honor of Murray Gell–Mann.

Spacetime Duality in String Theory*

John H. Schwarz

California Institute of Technology, Pasadena, CA 91125

Abstract

In string theory a compactified spatial dimension of radius R is equivalent to one of radius α'/R. This remarkable fact suggests that the Planck length, $\sqrt{\alpha'}$, is effectively the minimum distance that can be probed. Suitably generalized, it hints at a very large symmetry structure, which we attempt to explore by constructing appropriate low-energy effective Lagrangians. The enhanced gauge symmetry that occurs for $R = 1$ is shown to be spontaneously broken for $R \neq 1$. Suitably generalized, these features could give useful hints about the structure of string field theory.

Dedicated to Murray Gell-Mann
on the occasion of his 60th birthday

---

* This work supported in part by the U.S. Department of Energy under Contract No. DE-AC0381-ER40050

Introduction

Like many aspects of string theory, the use of the word "duality" has a curious history. Phenomenological studies of hadronic interactions in the 1960s led to the realization that high energy amplitudes could be approximated either by a sum of direct-channel resonance contributions or by a dual description in terms of crossed-channel Regge-pole exchanges. The desire to construct explicit mathematical examples was a central motivation of Veneziano and others in the original papers on string theory — called "dual resonance models" at the time [19]. Re-expressed more abstractly in the language of two-dimensional conformal field theory, this property was called the "conformal bootstrap" by Belavin, Polyakov, and Zamolodchikov [5]. Since this phenomenon is associated to the string world sheet, and to avoid confusion with other uses of the word duality, I propose referring to it as "world-sheet duality."

A second use of the word duality has arisen in the description of string theories with toroidally compactified spacetimes. When one spatial coordinate forms a circle of radius R the momentum component of the string in this dimension is quantized ($p = n/R$). In addition, since there is a nontrivial fundamental group, a closed string can wrap around the circle m times. Such an (m, n) string state has a "zero-mode" contribution to its mass–squared proportional to $n^2 a^2 + m^2/a^2$ where $a = \sqrt{\alpha'}/R$.[†] As we will discuss, various generalizations of the Z_2 duality symmetry of this solution have been found. Indeed, part of our purpose is to explore how much this duality symmetry can be extended. Clearly, this use of the word "duality" is quite different from the previous one. Since it concerns equivalences between seemingly distinct spacetime manifolds, I propose to call it "spacetime duality." This is the usage in the title of this paper. Admittedly, both types of duality have world-sheet and spacetime aspects, but I think this terminology makes a useful distinction.

Spacetime duality is not always a symmetry. Sometimes one solution at radius R is equivalent to a different one at radius $1/R$. (I find it convenient to set $\alpha' = 1$, even

---

† This formula was first given in ref. [16]. It was evident, but not emphasized, that this is invariant under the simultaneous interchanges $m \leftrightarrow n$ and $R \leftrightarrow \alpha'/R$. This fact was pointed out explicitly in ref. [20,26].

though the choice $\alpha' = 1/2$ is more common.) One significant example relates $E_8 \times E_8$ to spin(32)/Z_2 heterotic strings [27,12] and even non-supersymmetric ground states [13,18]. Another relates type IIA and type IIB superstrings [10]. Incidentally, this gives compelling evidence in each case that one is dealing with distinct solutions of a single theory rather than two different theories. (I know of no compelling argument, however, that the type II and heterotic theories should be identified.)

Clearly, spacetime duality is trying to tell us something important. It suggests that there is effectively a minimum distance — the Planck length — that can be probed in string theory [17,2]. This feature is expected in any sensible quantum theory of gravity, so it is pleasing to see it emerge. It also suggests a very amusing cosmological scenario if one imagines starting with a large contracting universe [7]. When the "radius" passes through the Planck length nothing very special happens except that gravity becomes strong, the temperature approaches the Hagedorn temperature, and the detailed dynamics becomes rather complicated. Later, however, when $R << 1$, this can be reinterpreted (thanks to spacetime duality) as an expanding universe. Thus one has the possibility of effectively having a "bounce," without requiring any exotic forces to stop the collapse! The validity of such a picture depends, among other things, on the existence of a meaningful definition of "time" when $R \approx 1$.

The partition function for strings at finite temperature involves calculating traces with a periodic Euclidean time of circumference β. Thus, as many authors have noted, one obtains expressions that are mathematically much the same as for a circular spatial dimension. (There are some differences, however, such as the periodicity properties of fermi fields.) In particular, there is an apparent $\beta \to 1/\beta$ duality. It is unclear what to make of this, however, since it seems unlikely that straightforward thermodynamic concepts apply near the Hagedorn temperature, where quantum gravity effects become strong. Also, the physical interpretation of temporal winding modes is confusing and controversial [28,21,3,6].[‡]

[‡] I am grateful to P. Schwarz for discussions of these issues.

A significant feature of spacetime duality is that there is often enhanced gauge symmetry at fixed points of the transformation. This is generically the case for non-supersymmetric sectors (bosonic strings and heterotic left-movers) but not for supersymmetric ones (Type II strings and heterotic right-movers). For example, in the case of circular compactification of a nonsupersymmetric dimension, the $U(1)$ gauge symmetry extends to $SU(2)$ at $R = 1$. The $U(1)$ has a simple Kaluza–Klein interpretation, but the $SU(2)$ is a stringy phenomenon. It seems reasonable to suppose that for $R \neq 1$, there is still $SU(2)$ gauge symmetry, but it is spontaneously broken [10]. This viewpoint is supported by constructing low-energy effective Lagrangians containing (in addition to the gauge fields) appropriate scalar fields that act as order parameters. When one dimension is compactified, the analysis is fairly straightforward. However, for tori of two or more dimensions, the requisite gauge symmetry is apparently so large that it must incorporate a substantial portion of the string theory. We have not pushed that analysis very far, but it may be an interesting avenue to pursue.

Toroidal Compactification

Consider d toroidally compactified string dimensions. They can be described by the two-dimensional string action

$$\frac{1}{2} \int d^2\sigma \left[\partial_\alpha X^i \partial^\alpha X^i + \epsilon^{\alpha\beta} B_{ij} \partial_\alpha X^i \partial_\beta X^j \right] \tag{1}$$

with the periodic identifications

$$X^i \approx X^i + 2\pi \sum_{a=1}^{d} e_a^i m^a, \tag{2}$$

where $\{m^a\}$ are arbitrary integers. This defines the toroidal configuration space as R^n modulo a lattice defined by the basis vectors $\{e_a^i\}$. The coefficients $B_{ij} = -B_{ji}$ are taken to be constants, which implies that the B term in the action is topological and does not modify the equation of motion $\partial_\alpha \partial^\alpha X^i = 0$. The world-sheet time is denoted by τ and the spatial coordinate by σ with period 2π.

A completely equivalent procedure is to define new string coordinates $X^a = e^a_i X^i$, where e^a_i is the inverse of the matrix e^i_a, and new B parameters $B_{ab} = e^i_a e^j_b B_{ij}$, as well as a metric tensor $G_{ab} = \sum_i e^i_a e^i_b$. In terms of these quantities, the action becomes [25]

$$\frac{1}{2} \int d^2\sigma \left[G_{ab} \partial_\alpha X^a \partial_\alpha X^b + \epsilon^{\alpha\beta} B_{ab} \partial_\alpha X^a \partial_\beta X^b \right]. \tag{3}$$

The periodicities in these variables are simply $X^a \approx X^a + 2\pi m^a$. In particular, if $X^a(2\pi, \tau) = X^a(0, \tau) + 2\pi m^a$, then m^a is called the winding number and X^a can be written as the sum of $m^a \sigma$ and a periodic term.

As we have already remarked, the B term in the action does not affect the equations of motion. However, it does affect the quantum theory as a consequence of its appearance in the canonical momentum

$$P_a(\sigma, \tau) = G_{ab} \dot{X}^b + B_{ab} X'^b, \tag{4}$$

where, as usual, $\dot{X} = \frac{\partial}{\partial \tau} X$ and $X' = \frac{\partial}{\partial \sigma} X$. Single-valuedness of the wave function on the torus implies that the zero modes of P_a are integers n_a. Thus in terms of the winding numbers m^a and discrete momenta n_a, the zero mode part of the X^a coordinate is given by

$$X^a_0 = x^a + m^a \sigma + G^{ab}(n_b - B_{bc} m^c)\tau, \tag{5}$$

where G^{ab} is the inverse of the matrix G_{ab}.

The parameters B_{ab} play a role analogous to the θ angle in QCD. Specifically, they are coefficients of a topological term in the action. As in the QCD case, this dependence is periodic. The dependence of the theory on each of the B_{ab} is periodic with unit period, since a shift $\Delta B_{ab} = N_{ab}$ for arbitrary integers N_{ab} can be compensated by a shift $N_{ab} m^b$ in the discrete momentum n_a. (One can check that the higher modes also allow this interpretation.) This provides equivalence relations on the space of theories parametrized by the moduli G_{ab}, B_{ab}. We will see that these are not the only ones.

Since $X^a(\sigma, \tau)$ satisfies a free wave equation, it has a "left–right" decomposition

$$X^a(\sigma, \tau) = X_L^a(\tau + \sigma) + X_R^a(\tau - \sigma). \tag{6}$$

These in turn have mode expansions of the form

$$X_L^a(\tau + \sigma) = x_L^a + p_L^a(\tau + \sigma) + \frac{i}{\sqrt{2}} \sum_{m \neq 0} \frac{\tilde{\alpha}_m^a}{m} e^{-im(\tau + \sigma)}$$

$$X_R^a(\tau - \sigma) = x_R^a + p_R^a(\tau - \sigma) + \frac{i}{\sqrt{2}} \sum_{m \neq 0} \frac{\alpha_m^a}{m} e^{-im(\tau - \sigma)}, \tag{7}$$

where $x^a = x_L^a + x_R^a$, and

$$p_L^a = \frac{1}{2}\left[m^a + G^{ab}(n_b - B_{bc} m^c)\right],$$

$$p_R^a = \frac{1}{2}\left[-m^a + G^{ab}(n_b - B_{bc} m^c)\right]. \tag{8}$$

Inserting X_R^a and X_L^a into the expression for the canonical momentum P_a gives its right- and left-moving pieces

$$P_{Ra} = (G - B)_{ab}\left[p_R^b + \frac{1}{\sqrt{2}} \sum_{m \neq 0} \alpha_m^b\, e^{-im(\tau - \sigma)}\right],$$

$$P_{La} = (G + B)_{ab}\left[p_L^b + \frac{1}{\sqrt{2}} \sum_{m \neq 0} \tilde{\alpha}_m^b\, e^{-im(\tau + \sigma)}\right]. \tag{9}$$

The standard commutation rules

$$[\alpha_m^i, \alpha_n^j] = [\tilde{\alpha}_m^i, \tilde{\alpha}_n^j] = m\delta_{m+n,0}\,\delta^{ij} \tag{10}$$

correspond to the relations

$$[\alpha_m^a, \alpha_n^b] = [\tilde{\alpha}_m^a, \tilde{\alpha}_n^b] = m\delta_{m+n,0} G^{ab} \tag{11}$$

We are dealing with a simple unconstrained quantum system. Thus straightforward canonical quantization gives

$$\left[X^a(\sigma,\tau), P_b(\sigma',\tau)\right] = -2\pi i \delta^a_b \, \delta(\sigma-\sigma') \ . \tag{12}$$

Let us compute $[X^a_L, P_{Lb}]$ and $[X^a_R, P_{Rb}]$ separately using the formulas above. One finds

$$\begin{aligned}\left[X^a_R(\sigma,\tau), P_{Rb}(\sigma',\tau)\right] &= -\frac{i}{2}(G+B)_{bc} \, G^{ac} \sum_{-\infty}^{\infty} e^{-im(\sigma-\sigma')} \\ &= -\pi i (1 - G^{-1}B)^a{}_b \, \delta(\sigma-\sigma')\end{aligned} \tag{13}$$

Similarly,

$$\left[X^a_L(\sigma,\tau), P_{Lb}(\sigma',\tau)\right] = -\pi i (1 + G^{-1}B)^a{}_b \, \delta(\sigma-\sigma') \tag{14}$$

so that the sum does have the desired value. It is interesting that the two expressions are unequal for $B \neq 0$.

The Hamiltonian, constructed in the usual manner, is

$$\mathcal{H} = \frac{1}{2} G_{ab} \left(\dot{X}^a \dot{X}^b + X^{a'} X^{b'} \right) \propto G_{ab} \left(X'^a_L X'^b_L + X'^a_R X'^b_R \right) \ . \tag{15}$$

The zero mode of this gives the mass-squared operator*

$$M^2 = G_{ab} \left[p^a_L p^b_L + p^a_R p^b_R + \sum_{m=1}^{\infty} \left(\alpha^a_{-m} \alpha^b_m + \tilde{\alpha}^a_{-m} \tilde{\alpha}^b_m \right) \right] \ . \tag{16}$$

This can be rewritten in the form

$$\begin{aligned}M^2 &= \frac{1}{2} G_{ab} \, m^a m^b + \frac{1}{2} G^{ab}(n_a - B_{ac}m^c)(n_b - B_{bd}m^d) \\ &+ \sum_{m=1}^{\infty} (\alpha^i_{-m} \alpha^i_m + \tilde{\alpha}^i_{-m} \tilde{\alpha}^i_m) \ .\end{aligned} \tag{17}$$

Note that the "background fields" G_{ab} and B_{ab} only appear in the zero mode contributions. When all degrees of freedom are included, the left- and right-moving

---

* Aside from an overall normal-ordering constant.

contributions to M^2 must agree (as a consequence of σ translational symmetry). The contribution of the zero modes to the difference is

$$G_{ab}(p_L^a p_L^b - p_R^a p_R^b) = m^a n_a . \tag{18}$$

It is fortunate that this is an integer, so that it can be balanced by oscillator contributions.

It is striking that the equation above is invariant under interchange of the winding numbers and discrete momenta $m^a \leftrightarrow n_a$. This turns out to be a symmetry of the entire spectrum (and even the interacting theory) provided that one simultaneously redefines B_{ab} and G_{ab} in an appropriate way. Rewriting the zero-mode contribution to M^2 in the form

$$\frac{1}{2}\Big[m^a(G - BG^{-1}B)_{ab}m^b + n_a G^{ab} n_b$$
$$+ m^a(BG^{-1})_a{}^b n_b - n_a(G^{-1}B)^a{}_b m^b\Big] . \tag{19}$$

It is clear that $m^a \leftrightarrow n_a$ must be accompanied by

$$(G - BG^{-1}B) \leftrightarrow G^{-1} \quad \text{and} \quad BG^{-1} \leftrightarrow -G^{-1}B . \tag{20}$$

This is possible to achieve since both of these equations are consequences of the single interchange [15, 29]

$$(G + B) \leftrightarrow (G + B)^{-1} . \tag{21}$$

By taking the transpose of this relation one also has $(G - B) \leftrightarrow (G - B)^{-1}$, and the conditions above are easily derived. Note that whereas $p_L \cdot p_L$ and $p_R \cdot p_R$ are preserved, there is no simple rule for $p_L \cdot p_R$ (contrary to some claims in the literature).

To understand the significance of the duality symmetry $(G + B) \leftrightarrow (G + B)^{-1}$, consider a single compactified dimension. If X has radius R, $e_1^1 = R$ and $G_{11} = R^2$. Thus the duality symmetry means that compactification on a circle of radius R is equivalent to compactification on a circle of radius $1/R$, provided winding numbers and discrete momenta are interchanged. Both are summed over all integers in forming the partition function, which therefore is invariant under $R \leftrightarrow 1/R$. In the general case it is invariant under $(G+B) \leftrightarrow (G+B)^{-1}$ as well as $B_{ab} \to B_{ab} + N_{ab}$, which we discussed earlier. The general duality symmetry implies that the Lorentzian lattice spanned by vectors $\sqrt{2}(p_L^a, p_R^a)$ with inner product

$$\sqrt{2}(p_L, p_R) \cdot \sqrt{2}(p_L', p_R') = 2G_{ab}(p_L^a p_L'^b - p_R^a p_R'^b) = (m^a n_a' + m'^a n_a) \qquad (22)$$

is even and self-dual.

It is interesting to explore the moduli space parametrized by G_{ab} and B_{ab} a little further. In the trivial one-dimensional case, after modding out by the Z_2 symmetry $G \leftrightarrow G^{-1}$, one is left with $G \geq 1$. (Orbifold constructions, which give additional interesting models, are not considered here.) In the case of a d-dimensional torus, the appearance of $p_L \cdot p_L$ and $p_R \cdot p_R$ in the M^2 formula and the left–right constraint show that locally the moduli space is $O(d,d)/O(d) \times O(d)$. This space, which has dimension d^2, is parametrized by G_{ab} and B_{ab} [25].

The global geometry is determined by the group of discrete symmetries generated by $B_{ab} \to B_{ab} + N_{ab}$ and $(G+B) \to (G+B)^{-1}$. Since these do not commute, they generate a rather large discrete group, which we would now like to determine. The principle that determines these symmetry transformations is that they preserve $p_L \cdot p_L \pm p_R \cdot p_R$. As we have seen this means they preserve

$$2m \cdot n = (m\ n) \begin{pmatrix} 0 & 1 \\ 1 & 0 \end{pmatrix} \begin{pmatrix} m \\ n \end{pmatrix} \qquad (23)$$

and

$$mGm + (n + mB)G^{-1}(n - Bm) = (m\ n) M \begin{pmatrix} m \\ n \end{pmatrix} \qquad (24)$$

where

$$M = \begin{pmatrix} G - B\,G^{-1}B & BG^{-1} \\ -G^{-1}B & G^{-1} \end{pmatrix}. \qquad (25)$$

In terms of $2d \times 2d$ matrices written in $d \times d$ blocks, we have

$$\begin{pmatrix} m \\ n \end{pmatrix} \to \begin{pmatrix} m' \\ n' \end{pmatrix} = \begin{pmatrix} \alpha & \beta \\ \gamma & \delta \end{pmatrix} \begin{pmatrix} m \\ n \end{pmatrix} = A \begin{pmatrix} m \\ n \end{pmatrix}, \tag{26}$$

where A has integral entries. Preservation of $2m \cdot n$ requires that

$$A^T \begin{pmatrix} 0 & 1 \\ 1 & 0 \end{pmatrix} A = \begin{pmatrix} 0 & 1 \\ 1 & 0 \end{pmatrix}, \tag{27}$$

which implies that $A \in O(d,d)$. The corresponding transformation of $G + B$ is given by

$$(m'\ n') M' \begin{pmatrix} m' \\ n' \end{pmatrix} = (m\ n) M \begin{pmatrix} m \\ n \end{pmatrix}. \tag{28}$$

In other words, $A^T M' A = M$. For example, $A = \begin{pmatrix} 0 & 1 \\ 1 & 0 \end{pmatrix}$ corresponds to $(G+B) \to (G+B)^{-1}$. Since $A \in O(d,d)$ has integral entries, it is tempting to conclude that the group of equivalences is $O(d,d;Z)$, but this is not quite right. $O(d,d)$ is conventionally defined as the group of matrices that preserve $\begin{pmatrix} 1 & 0 \\ 0 & -1 \end{pmatrix}$, which is related to the above by a change of basis

$$\eta^T \begin{pmatrix} 0 & 1 \\ 1 & 0 \end{pmatrix} \eta = \begin{pmatrix} 1 & 0 \\ 0 & -1 \end{pmatrix}, \tag{29}$$

where

$$\eta = \frac{1}{\sqrt{2}} \begin{pmatrix} 1 & 1 \\ 1 & -1 \end{pmatrix}. \tag{30}$$

In this basis

$$A \to \tilde{A} = \eta^T A \eta = \frac{1}{2} \begin{pmatrix} \alpha + \beta + \gamma + \delta & \alpha + \gamma - \beta - \delta \\ \alpha + \beta - \gamma - \delta & \alpha + \delta - \beta - \gamma \end{pmatrix}. \tag{31}$$

The entries of this matrix can be half integers as well as integers. Thus the group in question is not quite the same as $O(d,d;Z)$. We denote it by $\tilde{O}(d,d;Z)$. It is quite analogous to the symplectic modular group — the integral matrices that preserve $\begin{pmatrix} 0 & 1 \\ -1 & 0 \end{pmatrix}$. The shift $B_{ab} \to B_{ab} + N_{ab}$, with N an antisymmetric matrix of integers, corresponds to

$$A = \begin{pmatrix} 1 & 0 \\ -N & 1 \end{pmatrix} \text{ or } \tilde{A} = \begin{pmatrix} 1 - \frac{N}{2} & \frac{N}{2} \\ -\frac{N}{2} & 1 + \frac{N}{2} \end{pmatrix}. \tag{32}$$

Note that $\tilde{A}$ would only be integral for even integers N_{ab}.

Altogether we conclude that the moduli space of inequivalent toroidal compactifications of d dimensions, parametrized by inequivalent (G, B) is $O(d,d)/O(d) \times O(d)$ with real coefficients. Its global structure is specified by also modding out by $\tilde{O}(d,d;Z)$, as defined above. This group action has fixed points, which result in orbifold points of the resulting moduli space. As we will explore, these points correspond to enhanced gauge symmetry.

An additional remark that should be made is that the enhanced discrete symmetry generated by integer shifts of B has nontrivial consequences even for theories with $B = 0$. Specifically, one can determine the subgroup of $\tilde{O}(d,d;Z)$ that preserves matrices of the form $M = \begin{pmatrix} G & 0 \\ 0 & G^{-1} \end{pmatrix}$. The general solution is rather complicated, but it certainly contains as a subgroup matrices of the form

$$A = \begin{pmatrix} a & 0 \\ 0 & (a^{-1})^T \end{pmatrix} \tag{33}$$

with $a \in SL(d, Z)$. This corresponds to the transformation $G \to a^T G a$. Adjoining the duality transformation $G \to G^{-1}$, extends the group by a Z_2 factor. Another factor of 2 corresponds to matrices with $\det a = -1$.

Let us consider some of the vector particles arising from toroidal compactification. One always obtains Kaluza–Klein $U(1)$ factors associated with the toroidal isometries. In terms of oscillators the associated massless vector states are represented by

$$\zeta \cdot \alpha_{-1} \, \tilde{\alpha}^i_{-1} |0\rangle \quad , \quad \zeta \cdot \tilde{\alpha}_{-1} \, \alpha^i_{-1} |0\rangle \; . \tag{34}$$

Thus there is a $[U(1)]^d$ factor associated with left-movers and another $[U(1)]^d$ associated with right-movers. (In the heterotic theory only the left-moving factor would appear.) Additional left-moving vector particles can be constructed using winding number and internal momentum excitations

$$\zeta \cdot \alpha_{-1} \, e^{i(p_L \cdot x_L + p_R \cdot x_R)} |0\rangle \quad . \tag{35}$$

Since

$$p_L^2 = \frac{1}{4}\Big[mGm + (n + mB)G^{-1}(n - Bm) + 2m \cdot n\Big] \, ,$$
$$p_R^2 = \frac{1}{4}\Big[mGm + (n + mB)G^{-1}(n - Bm) - 2m \cdot n\Big] \, , \tag{36}$$

level-matching is achieved for $m \cdot n = 1$. This gives an infinite spectrum of vector states with masses

$$M_V^2 = mGm + (n + mB)G^{-1}(n - Bm) - 2 \quad . \tag{37}$$

It is easy to prove that $M_V^2 \geq 0$ for all choices of (m, n) satisfying $m \cdot n = 1$. One method is to write it as a square $|mG^{1/2} - nG^{-1/2} - mBG^{-1/2}|^2$. Right-moving vector states can be constructed in an analogous manner for $m \cdot n = -1$.

We have asserted that there is enhanced gauge symmetry at fixed points of the $\tilde{O}(d, d; Z)$ transformation group. A very simple example is $G = 1, B = 0$, which is a fixed point of $(G + B) \to (G + B)^{-1}$. $G = 1$ corresponds to all d compactified dimensions having radius $R = 1$ (for our choice of units) and being orthogonal. In this case

$$M_V^2 = m \cdot m + n \cdot n - 2 \quad . \tag{38}$$

This vanishes for $2d$ choices of (m, n) satisfying $m \cdot n = 1$. Specifically $m = n$ and one component is ± 1 while the others vanish. These $2d$ massless vector particles extend the $[U(1)]^d$ gauge symmetry to $[SU(2)]^d$, a fact that is well-known.

The spectrum of scalar particles can be explored in similar fashion. Those without oscillator excitations are given by

$$e^{i(p_L \cdot x_L + p_R \cdot x_R)} |0\rangle , \qquad (39)$$

where now level-matching requires $m \cdot n = 0$. In this case the masses are given by

$$M_S^2 = mGm + (n + mB)G^{-1}(n - mB) - 4 . \qquad (40)$$

The state with $m = n = 0$ corresponds to the standard tachyonic ground state. Whether or not the spectrum contains additional tachyons depends on the values of the G and B background fields. For example, in the $[SU(2)]^d$ case one has scalar masses of the form $M_S^2 = m \cdot m - 4$.

Effective Field Theories

Let us examine the significance of some of the results we have obtained in more field theoretic terms. This is done by writing down an effective field theory containing fields associated with those modes we wish to examine. Other modes are ignored. This means that they are assumed to be integrated out (at tree level) if they are massive or simply set equal to trivial flat space values if they are massless. The latter case applies to the graviton, dilaton, and antisymmetric tensor, in particular.

The simplest case to consider is the bosonic string with one circular dimension of radius R. The relevant vector particles are massless ones corresponding to unbroken $U(1) \times U(1)$ gauge symmetry and four additional ones with $U(1)$ charges $(\pm 1, 0)$ and $(0, \pm 1)$ and $M^2 = R^{-2} + R^2 - 2$. At the self-dual point these combine with the $U(1)$'s to give $SU(2) \times SU(2)$ symmetry. Our goal is to write an $SU(2) \times SU(2)$ gauge theory that is spontaneously broken when a certain scalar field takes a value corresponding to $R \neq 1$.

To carry out the program described above, it is clear that certain scalar fields should be included in our effective field theory. The ones we will keep are those that become massless at the self-dual point $R = 1$. One of them, given by the Fock space state $\alpha_{-1}\tilde{\alpha}_{-1}|0\rangle$ (where α_{-1} and $\tilde{\alpha}_{-1}$ refer to the compact dimension) has $(I_3, I_3') = (0,0)$ and is massless for all R. Four others have the same (I_3, I_3') quantum numbers and masses as the charged vector states. They are obtained from the corresponding vector states by the substitution $\zeta_\mu \alpha_{-1}^\mu \to \alpha_{-1}$. These are the Goldstone bosons that get "eaten" for $R \neq 1$. Finally, there are two states with $(I_3, I_3') = \pm(1,1)$ and $M^2 = 4R^{-2} - 4$ and two with $(I_3, I_3') = \pm(1,-1)$ and $M^2 = 4R^2 - 4$. Altogether, at the self-dual point these nine scalars form an $(I, I') = (1,1)$ massless multiplet of $SU(2) \times SU(2)$. For $R \neq 1$, four are eaten by vector states and the other five have the masses indicated above.

Let us look more closely at the effective field theory. The $(1,1)$ scalars enumerated above are described by a field ϕ_{ab}, where a and b are adjoint indices for the two $SU(2)$ factors. Our goal is to construct a suitable potential $V(\phi)$ that will give rise to the Higgsism and mass spectrum that we want. The key identification that will give the desired results is

$$e^{\langle \phi_{33} \rangle} = 1/R^2 \ . \tag{41}$$

For this choice the duality transformation corresponds to $\phi_{33} \to -\phi_{33}$, which is part of a finite gauge transformation.

Evidence for the interpretation of ϕ_{33} given above can be provided by considering certain additional scalars that have been ignored until now. Namely, there are four states with isospin quantum numbers $(\frac{1}{2}, \frac{1}{2})$ whose mass squared at the symmetric point is -3. Two have $m = \pm 1, n = 0$ and $M^2 = R^{-2} - 4$ and the other two have $m = 0, n = \pm 1$ and $M^2 = R^2 - 4$. Let us represent these scalars by a bispinor ψ. Then their mass values can be described by an $SU(2) \times SU(2)$ symmetric interaction with the ϕ_{ab} multiplet as follows: First define a 4×4 matrix $\Sigma = \phi_{ab} \tau^a \otimes \sigma^b$, where τ and σ are Pauli matrices. Then $V = \psi^\dagger (e^\Sigma - 4)\psi$ is easily seen to be an $SU(2) \times SU(2)$ invariant term that gives the desired pattern of masses.

Let us now turn to the more subtle matter of ϕ self-couplings.* We require an $SU(2) \times SU(2)$ invariant potential that gives the appropriate mass spectrum as a function of $\langle\phi_{33}\rangle$. The first thing to note is what those masses are. The states with $(I_3, I_3') = \pm(1,1)$ should have $M^2 = 4R^{-2} - 4$ and those with $(I_3, I_3') = \pm(1,-1)$ should have $M^2 = 4R^2 - 4$. Introducing $SU(2)$ indices $\pm = (1 \pm i2)$, it follows that V should contain

$$V_{\text{mass}} = (e^{\phi_{33}} - 1)\phi_{++}\phi_{--} + (e^{-\phi_{33}} - 1)\phi_{+-}\phi_{-+} . \tag{42}$$

The $(0,0)$ state (corresponding to ϕ_{33}, itself) should be massless for all $\langle\phi_{33}\rangle$. The $\phi_{3\pm}$ and $\phi_{\pm 3}$ fields correspond to Goldstone bosons that are eaten by vector states for $\langle\phi_{33}\rangle \neq 0$. Therefore they should not have mass terms. The part of the answer written above needs to be extended to an $SU(2) \times SU(2)$ invariant expression. This can be examined as a power series in ϕ. The leading terms are cubic and contain $\phi_{33}(\phi_{++}\phi_{--} - \phi_{+-}\phi_{-+})$. This is (uniquely) achieved by

$$V_3 \propto \det \phi = \epsilon_{abc}\, \epsilon_{a'b'c'}\, \phi_{aa'}\, \phi_{bb'}\, \phi_{cc'} . \tag{43}$$

At the quartic level there are two possible invariants:

$$V_4^{(1)} = (\text{tr } \phi^T \phi)^2 , \quad V_4^{(2)} = \text{tr } \phi^T \phi \phi^T \phi . \tag{44}$$

The requirement that there be no mass terms for ϕ_{3a} and ϕ_{a3} implies that $V_4 \propto V_4^{(1)} - V_4^{(2)}$. The correct coefficient can be determined by matching to the quartic piece of V_{mass}. As we continue to higher powers of ϕ the number of possible invariants increases, but the number of conditions does not. Thus V is not uniquely determined by the masses alone. One can do better by examining the self-interactions in the full string theory, but that is a very involved process. Let us instead move on to the next problem — the compactification of more than one dimension.

---

* The following construction was carried out in collaboration with M. Douglas.

For compactification on a torus of more than one dimension interesting new issues arise. These are adequately illustrated by considering a two-dimensional torus $S^1 \times S^1$. As we have seen earlier, this case always has $U(1)^2$ gauge symmetry for both the left and the right movers ($U(1)^4$, altogether). Moreover, for the special case $G = 1$, $B = 0$, each $U(1)$ factor is extended to $SU(2)$. The new feature is that there is another point of enhanced symmetry in which $U(1)^2 \to SU(3)$. Thus there are two distinct maximal symmetry groups $SU(2) \times SU(2)$ and $SU(3)$, neither of which is a subgroup of the other. Also, there is no point in moduli space that gives a larger symmetry group containing both of these as a subgroup.

If we wish to mimic the effective action analysis given above for S^1 compactification, there are various strategies one might consider. The simplest one is to choose one of the two enhanced symmetry points and construct an effective action based on gauge fields for that symmetry and the scalars that are massless at the symmetry point. Such an analysis is a straightforward analog of the one discussed earlier. It doesn't really teach us anything new, however. Its defect is that whichever group we choose to display, the enhanced symmetry at other special points is not going to appear. The fields in the action are only sufficient to exhibit an $SU(2) \times U(1)$ subgroup.

It would be nice if we could construct an effective action containing scalar fields whose expectation values determine the four order parameters $G + B$ in such a way that the appropriate gauge symmetry would appear throughout the moduli space. Also, the discrete symmetries $\tilde{O}(2,2;Z)$ should be identifiable as finite gauge transformations. Such a setup undoubtedly is realized in the full string theory, the question is how much of a truncation is possible while maintaining these features. In the S^1 case we saw that a quite simple system would suffice. Now, however, we require gauge fields for both $SU(3)$ and $SU(2) \times SU(2)$ (doubled when both left- and right-movers are included). This raises the question: what is the minimal group that contains both of these as subgroups? Whatever it is, it will never occur as an unbroken symmetry, and it could be quite exotic. There is no solution with rank two. It appears that the answer requires including an infinite number of the vector states of the string theory spectrum – maybe even all. At this point the restriction to scalars and vectors seems very artificial. Perhaps the desired features can only be properly realized in the full-fledged string field theory, which has not yet been formulated, at least in a suitable form. Maybe, the issues we have raised will give some hints about essential features of its structure.

Conclusion

The moduli space of possible string theory compactifications, characterized by the expectation values of scalar fields, has a rich structure. Special points with enhanced gauge symmetry are fixed points of discrete transformations that relate equivalent compactifications. Viewing the string theory as a gauge theory, the discrete transformations can be interpreted as finite gauge transformations. These features have been exhibited explicitly in a simple field theory model for the case of circular compactification. The study of more complicated cases may be helpful for developing an understanding of string field theory.

I am grateful to M. Douglas and A. Shapere for helpful discussions.

REFERENCES

1. E. Alvarez and M. A. R. Osorio, "Duality is an Exact Symmetry of String Perturbation Theory," Madrid preprint FTUAM/89-06 (1989).

2. D. Amati, M. Ciafaloni, and G. Veneziano, Phys. Lett. **216B** (1989) 41.

3. J. J. Atick and E. Witten, Nucl. Phys. **B310** (1988) 291.

4. M. J. Bowick and S. B. Giddings, "High-Temperature Strings," Harvard preprint HUTP-89/A007 (1989).

5. A. A. Belavin, A. M. Polyakov, and A. B. Zamolodchikov, Nucl. Phys. **B241** (1984) 333.

6. M. J. Bowick and S. B. Giddings, "High-Temperature Strings," Harvard preprint HUTP-89/A007 (1989).

7. R. Brandenberger and C. Vafa, "Superstrings in the Early Universe," Brown preprint HET-673 (1988).

8. L. Caneschi, Phys. Lett. **228B** (1988) 332.

9. R. Dijkgraaf, E. Verlinde, and H. Verlinde, Commun. Math. Phys. **115** (1988) 649.

10. M. Dine, P. Huet, and N. Seiberg, "Large and Small Radius in String Theory," preprint IASSNS-HEP-88/54 and CCNY-HEP-88/20 (1988).

11. S. Ferrara, D. Lüst, A. Shapere, and S. Theisen, Phys. Lett. **225B** (1989) 363.

12. P. Ginsparg, Phys. Rev. **D35** (1987) 648.

13. P. Ginsparg and C. Vafa, Nucl. Phys. **B289** (1987) 414.

14. P. Ginsparg, Nucl. Phys. **B295** (1988) 153.

15. A. Giveon, E. Rabinovici, and G. Veneziano, Nucl. Phys. **B322** (1989) 167.

16. M. B. Green, J. H. Schwarz, and L. Brink, Nucl. Phys. **B198** (1982) 474.

17. D. J. Gross, "Superstrings and Unification, "Princeton preprint PUPT-1108 (1988).

18. H. Itoyama and T. R. Taylor, Phys. Lett. **186B** (1987) 129.

19. M. Jacob, editor, "Dual Theory," Physics Reports Reprint Volume I (North-Holland, 1974).

20. K. Kikkawa and M. Yamasaki, Phys. Lett. **149B** (1984) 357.

21. Ya. I. Kogan, JETP Lett. **45** (1987) 709.

22. J. Lauer, J. Mas, and H. P. Nilles, Phys. Lett. **226B** (1989) 251.

23. W. Lerche, D. Lüst, and N. P. Warner, "Duality Symmetries in N=2 Landau-Ginzburg Models," CERN-TH.5504/89 and CALT-68-1575 (1989).

24. V. P. Nair, A. Shapere, A. Strominger, and F. Wilczek, Nucl. Phys. **B287** (1987) 402.

25. K. S. Narain, Phys. Lett. **169B** (1986) 41.

26. N. Sakai and I. Senda, Prog. Theor. Phys. **75** (1984) 692.

27. K. S. Narain, M. H. Sarmadi, and E. Witten, Nucl. Phys. **B279** (1987) 369.

28. B. Sathiapalan, Phys. Rev. Lett. **58** (1987) 1597.

29. A. Shapere and F. Wilczek, Nucl. Phys. **B320** (1989) 669.

30. C. Vafa, "Quantum Symmetries of String Vacua," Harvard preprint HUTP-89/A021 (1989).

Supersymmetry and Quasi-Supersymmetry

Y. Nambu

Enrico Fermi Institute and Department of Physics
University of Chicago, Chicago, Illinois 60637

I first met Murray Gell-Mann when he popped up in my office at the Institute for Advanced Study, and described to me the isospin-strangeness rule he had discovered. He pronounced my name correctly and interpreted its meaning correctly. That was September 1953.

The post-War decades have been the Golden Age of particle physics. Theory and experiment went hand in hand to make amazing advances. What we know now about the world of elementary particles is incredibly richer than what we did forty years ago, and we owe this to Murray above all. Looking beyond the Baroque period we are in now, I hope Murray's spirit will come back alive again.

Recently I have been taking a renewed interest in the BCS mechanism as a model for spontaneous generation of fermion mass and associated Goldstone (G) and Higgs (H) collective bosons. Here I mean by a BCS mechanism the formation of fermion pair condensates due to a short range attraction. In an idealized situation, this may be represented by a four-fermion interaction, and the dynamics is essentially determined by the properties of fermion bubble diagrams. A characteristic feature of the bubble approximation is that the Bogoliubov–Valatin (BV) fermion and the Higgs boson have the simple mass ratio 1:2. Such modes are known to exist in superconductors.

These low energy modes can be represented by an effective Ginzburg–Landau–Gell-Mann–Lévy Hamiltonian in which the boson self-coupling and the boson-fermion Yukawa coupling are related so as to satisfy the mass ratios. It has been found, moreover, that the static part H_{st} of the Hamiltonian can be factored as an anticommutator of fermionic operators as[1]

$$H_{st} = \{\bar{Q}, \bar{Q}^+\}/2, \tag{1}$$

much like in supersymmetry. In the case of BCS superconductivity, one has

$$\begin{aligned}
\bar{Q} &= \int dv Q, \quad \bar{Q}^+ = \int dv Q^+, \\
Q &= \pi^+ \psi_{up} + W \psi_{dn}^+, \\
Q^+ &= \pi \psi_{up}^+ + W^+ \psi_{dn}, \\
W &= iG(\phi^+ \phi - c^2), \\
\phi &= h + ig, \pi = \pi_h + i\pi_g, \\
H_{st} &= \int dv[|\pi|^2 + |W|^2 + G(\phi^+ \psi_{up}\psi_{dn} + \phi \psi_{dn}^+ \psi_{up}^+)].
\end{aligned} \tag{2}$$

The fields of Goldstone (g) and Higgs (h) modes are here combined into a complex field ϕ, and their canonical conjugates π_g and π_h into a field π. The Higgs potential is represented as $|W|^2$. When h is expanded around c, the fermions and bosons acquire the correct mass values.

The situation is similar to the case of supersymmetric quantum mechanics, but the difference lies in the form of W. Had one chosen $W = G(\phi^2 + m\phi)$, one would have gotten an analog of the Wess-Zumino model. Here, however, the fermions ψ and ψ^+ carry electric charge ± 1, the bosons ϕ and ϕ^+ charge ± 2, while Q and Q^+ are constructed in such a way as to have charge ± 1 so that charge is conserved and H is neutral. As a consequence, one pays the price of the Q's not being nilpotent, so supersymmetry is intrinsically broken. Nevertheless, the bilinear parts of Q, Q^+ and H, obtained by expansion around c, form a spectrum generating superalgebra su $(1/2)$ with central extension, expressed in terms of the creation and annihilation operators for the BV-fermion and the H-boson. The Goldstone mode π_G, lacking its conjugate counterpart, serves as a dynamical central charge.

It is not clear why the BCS mechanism (in the bubble approximation) should have a built-in quasi-supersymmetry, but the above factorization works in general for multi-component fermion and boson fields. I used the word quasi-supersymmetry to characterize these theories.

It is tempting to seek relativistic versions of quasi-supersymmetry. The BCS picture fits well with hadron physics, where the quarks (and nucleons) are the BV-fermions, and the σ and π mesons are the composite H and G bosons. It also gives an attractive interpretation of the Higgs field in the Weinberg–Salam electroweak theory, in the same spirit as the technicolor theory. In technicolor models, the Higgs field is a composite of heavy technifermions, but recently I proposed a model in which the Higgs field is a composite one made of the ordinary fermions, primarily the t and $\bar{t}$.[2,3]

The reasoning that led to the model is twofold. One is the quasi-supersymmetry with its simple mass relations already present in the Nambu–Jona-Lasinio model. The other is the possibility of a bootstrap symmetry breaking: the scalar Higgs boson which is a result of a chiral symmetry breaking, may itself be responsible for supplying the attractive force necessary for the symmetry breaking. This would then offer a bare minimum electroweak unification with only the known fermion and gauge fields as elementary fields.

The BCS quasi-supersymmetry predicts a 2:1 ratio between the Higgs and the top mass, subject of course to renormalization corrections. The bootstrap condition was formulated by requiring the cancellation of quadratic divergences in the Higgs sector among the fermion, Higgs, and gauge field contributions, on the reasoning that the Weinberg–Salam theory, even if it was the effective low-energy limit of a more fundamental theory without the Higgs, should not depend on the underlying high energy scale through a quadratic cut-off. The absence of quadratic divergences is also a feature characteristic of supersymmetric theories.

This condition leads to a quadratic mass sum rule among the top, Higgs, W and

Z. Combining it with the top-Higgs mass ratio, one should then be able to get m_t and m_H in terms of the known m_W and m_Z. Unfortunately it has turned out that the two relations are compatible only if $m_W = m_Z = 0$, i.e., the top and Higgs bootstrap without the gauge fields.[2] But a modified set of rules, requiring the masses to be self-consistently generated by tadpoles and be free of quadratic divergences, leads to meaningful solutions.[3] Bardeen and Hill have proposed a scheme[3] in which the quadratic condition is discarded in favor of the picture that the Higgs and Higgs potential are the result of integrating out the high-momentum components of the fermion field, to which the low-momentum part of the field couples.

In the following I would like to seek possible dynamical bases for supersymmetry or some relaxed forms of supersymmetry, in the hope that they may shed light on the fermion mass matrix and the nature of the Higgs.

The fermionic charges Q and Q^+ in supersymmetry are spatial integrals of the time component of local supercurrents $Q^\mu(x), Q^{+\mu}(x)$. They are typically made up of the following pieces

$$Q_R^\mu(x) = \partial_\nu \phi(x) \sigma^\nu \tilde{\sigma}^\mu \psi_R(x) ,$$
$$Q_R^\mu(x) = F_{\lambda\rho}(x) \sigma^\lambda \tilde{\sigma}^\rho \sigma^\mu \psi_L(x) , \qquad (3)$$
$$Q_R^\mu(x) = W(x) \sigma^\mu \psi_L(x) ,$$

and similar forms with L and R interchanged. Here the time derivative $\partial_0 \phi$ is to be regarded as the canonical conjugate to ϕ. The gauge field is quantized in the temporal gauge.

In general these currents consist of spinors behaving like (0,1/2), (1/2,0), (1,1/2), and (1/2,1). One may assume that this would be the case even if supersymmetry is relaxed. In the BCS type examples, the Q and Q^+ operators carry internal quantum numbers, so one must look at extended supersymmetry, which reads

$$\{Q^i_{\alpha R}, Q^{+j}_{\beta R}\} = 2P_\mu(\tilde{\sigma}^\mu)_{\alpha\beta}\delta^{ij}, \quad \{Q^i_{\alpha R}, Q^j_{\beta R}\} = \epsilon_{\alpha\beta} Z^{ij}, (i,j=1,...N) . \qquad (4)$$

If supersymmetry is broken, the right-hand side of Eq. (4) will acquire additional operator terms. These terms are not conserved quantities in order to evade the no-go theorem. Being an integral of local densities but not conserved, they will not have simple tensor properties. In essence one is thus led to supercurrent algebras, or else one might consider operator product expansion of supercurrents. I will not pursue this line here, though.

Instead let me work with concrete examples to see what can be learned. In fact one immediately begins to appreciate the difficulty of beating the no-go theorem. As a starter, try to emulate the Wess–Zumino model by replacing the ψ_{up} and ψ_{dn}^+ of Eq. (2) with a two-component Weyl spinor ψ and its charge conjugate ψ^c, and adding a kinetic term:

$$Q = (\pi^+ + \sigma \cdot \nabla \phi^+)\psi + W\psi^c, \{\bar{Q}, \bar{Q}^+\}/2 = H ,$$
$$H = |\pi|^2 + |\nabla\phi|^2 + |W|^2 - i\psi^+\sigma \cdot \nabla\psi \qquad (5)$$
$$+ G(\psi^{c+}\psi\phi + \psi^+\psi^c\phi)/2 .$$

This Hamiltonian gives a Majorana mass $m_f = Gc$ for the fermion, and the standard Goldstone and Higgs masses $m_G = 0$, $m_H = 2Gc$, in the correct BCS ratios. One finds, however, the rest of the supersymmetry relations for the momentum $\{Q, \sigma^i Q\} = P^i$, and those involving the central charge, $\{Q, Q\}, \{Q^+, Q^+\}$, are not satisfied.

By working out the anticommutators explicitly, one learns two necessary conditions for Q and Q^+ to yield the Poincaré generators. One is that the fermion and boson degrees of freedom must match, because the fermion kinetic energy piece in H comes weighted by the number of boson field contractions, and the boson kinetic energy piece comes weighted by the number of fermion field contractions. The second and less trivial condition is that the momentum part of the anticommutator, $\{Q, \sigma^i Q^+\} = P^i$, should be purely kinematic, i.e., does not develop interaction terms. In supersymmetry, the potentials are such that these terms vanish either by being a total divergence, by symmetry, or by cancellation between different contributions.

Some realistic examples of dimension-matching were considered before.[2] But they in general fail to satisfy the second condition.

I will digress a bit here. For a dynamical analysis of supersymmetry, it is instructive to regard the usual anticommuting parameters as a dynamical quantity. Imagine an enlarged Hilbert space which includes the old Fermi and Bose fields that make up a chiral supermultiplet, as well as a new fermi field $\theta(x)$ and $\theta^+(x)$, and consider the following expression which couples ψ and θ:

$$J^\lambda = \theta_L^+(x)\sigma^\lambda \psi_L(x) + \psi^+(x)_L \sigma^\lambda \theta_L(x) . \tag{6}$$

J^λ is a conserved current if both fields are massless. Suppose, however, that θ has developed a Majorana mass as a result of symmetry breaking. Then J^λ is no longer conserved, and there will be an induced current coupled to a Goldstone boson ϕ:

$$I = J^\lambda \partial_\lambda \phi + h.c. . \tag{7}$$

Because of the identity

$$\sigma_\mu \tilde{\sigma}^\lambda \sigma^\mu = -2\sigma^\lambda , \tag{8}$$

this can be rewritten as

$$I = (-1/2)\theta_L^+ \sigma_\mu \tilde{\sigma}^\lambda \sigma_L^\mu \psi \partial_\lambda \phi = -(1/2)\theta_L^+ \sigma_\mu Q_L^\mu + h.c. . \tag{9}$$

In other words, supercurrents naturally appear in induced Goldstone terms when some of the Fermi fields split off from the rest by acquiring a mass. One might expect, in a similar manner, that the supercurrent for the gauge field, Eq. (3b), is related to an anomalous Pauli coupling induced by the symmetry breaking, but this does not come out naturally because $\tilde{\sigma}_\mu \sigma_\lambda \tilde{\sigma}_\rho F^{\lambda\rho} \sigma^\mu = 0$.

It is, however, possible to interpret θ as the spin 1/2 piece of a reducible Rarita–Schwinger field Θ_μ:

$$\begin{aligned}\Theta_{\mu R} &= \theta_{\mu R} + \sigma_\mu \theta_L , \\ \theta_L &= \tilde{\sigma}^\lambda \Theta_{\lambda R}/4 ,\end{aligned} \tag{10}$$

θ_μ being the gravitino field. Then Eq. (10) is the spin 1/2 part of the coupling

$$I' = \Theta_R^{+\mu} Q_{\mu L} + h.c. , \tag{11}$$

between Θ_μ and a composite Rarita–Schwinger field Q_μ, both of which are divergenceless. Of course Q_μ itself is not massless, but obeys an equation of motion

$$\sigma^\nu \partial_\nu Q^\mu = \partial_\lambda Q^{\mu\lambda} ,$$
$$Q^{\mu\lambda} = \phi(\tilde\sigma^\mu \partial^\lambda - \tilde\sigma^\lambda \partial^\mu)\psi. \tag{12}$$

These observations suggest that there may be some connection between Supersymmetry and symmetry breaking: supersymmetry characterizes the massless sector in the spontaneous symmetry breaking of a class of larger theories, in which the scalars arise as composite fields equipped with a special kind of Higgs potentials. The quasi-supersymmetry, such as is found in the BCS mechanism, might then belong to a somewhat more general class of theories. The improved divergence properties of supersymmetric theories seem to imply that the heavy particle sector is decoupled to a better degree than is normal. It may not be unreasonable to expect that a similar situation may happen in a more general class of theories.

I will now come back to the quasi-supersymmetry of the earlier sections. Its main motive was to find realistic models of unified gauge theories with no obvious supermultiplet structure. In view of the nonrelativistic prototypes, one would have to assume the fermionic charges to carry internal quantum numbers. As was mentioned already, however, relaxation of supersymmetry generates extra terms in Eq. (4) which do not have simple transformation properties.

The BCS mechanism involves only fermions and collective bosons as low energy modes. Elementary gauge fields may be responsible for the symmetry breaking, but in the Weinberg–Salam theory, the electroweak gauge fields play no dynamical roles. In supersymmetry, on the other hand, gauge couplings also contribute to the Yukawa couplings of the scalar field, and thereby control the fermion masses. This could also happen in the BCS mechanism, although it is not clear how.

Taking all these things into consideration, I will define relativistic quasi-supersymmetry to mean a theory that meets the following criteria:

1. The sum over internal indices $\sum_i \{Q^{+i}, Q^i\}$ yields the correct Poincare algebra. It implies in particular that the bosonic and fermionic degrees of freedom must match.

2. The constraints on the other anticommutators $\{Q, Q\}$ and (Q^+, Q^+) are ignored. This seems reasonable since the nilpotency of Q and Q^+ is no longer expected. Moreover, if they carry internal quantum numbers, one cannot in general form singlets out of two Q's or two Q^+'s, especially in models that resemble the real world.

Basically what is done here is to relax the rigid constraints of the extended supersymmetry, requiring only that the total energy-momentum of the system comes out as a sum of its parts. The meaning of the label i is thus somewhat different from that in extended supersymmetry.

I close with a concrete example as an existence proof. The model has two generations of SU(2) × U(1) fermion doublets ψ_1 and ψ_2, two Higgs doublets ϕ_1 and ϕ_2, and the SU(2) × U(1) gauges fields $F_i, i = 0, .., 3$. There are two Q's, Q_1 and Q_2, each being a SU(2) doublet.

$$Q_1 = [(\pi + \sigma \cdot D\phi)\psi_R + (\sigma \cdot F_+ + W)\psi_L]_1 ,$$
$$Q_2 = [(\pi + \sigma \cdot D\phi)\psi_R + (\sigma \cdot F_+ - W)\psi_L]_2 ,$$
$$\phi = \phi_0 + i\tau_i\phi_i, \quad \pi = \pi_0 + i\tau_i\phi_i ,$$
$$F_+ = F_{+0} + \tau_i F_{+i}, F_+ = B + iE, W = W_1 - W_2 , \qquad (13)$$
$$W_1 = G_1(\phi_1 h_1 \phi_1^+ - c_1^2), h_1 = a_1 + b_1\tau_3 ,$$
$$W_2 = G_2(\phi_2 h_2 \phi_2^+ - c_2^2), h_2 = a_2 + b_2\tau_3 ,$$
$$D\phi = \partial\phi - i(g/2)\tau_i A_i\phi - i(g'/2)\phi\tau_3 A_0 .$$

The matrices h control how each Higgs doublet couples to the right-handed up and down fermions, but it will be seen that they are determined by the gauge couplings.

As the Q's carry gauged quantum numbers, the computation of the anticommutators of local densities requires some care. One has to insert Wilson lines between the Q's in order to get the correct covariant derivatives.

When the sum $\sum\{Q, Q^+\}$ over the internal indices is formed, one has to make sure that interaction terms are absent in the momentum part of the algebra. They arise as interference terms among the three components, Eq. (3), that make up the Q's. First, the W terms are so designed that the interference between F and W vanishes when summed over Q_1 and Q_2. The other two interference terms, $\phi - F$ and $\phi - W$, give rise to Yukawa couplings. The $\phi - F$ contribution has different weights, 3 versus -1, for energy and momentum. The condition for the absence of Yukawa terms in the momentum thus reads

$$(2Ga)_k = 3g/2, \quad (6Gb)_k = g'/2 \ (k = 1, 2), \tag{14}$$

giving the relative Yukawa coupling strengths (in the energy) to the right-handed fermions,

$$\begin{aligned} 3g + g' \text{ for up}, \\ 3g - g' \text{ for down} \end{aligned} \tag{15}$$

for both fermion generations. The SU(2) coupling g is common to both generations, but g'_1 and g'_2 may be different.

I am indebted to Drs. W.A. Bardeen, C.T. Hill, and M. Ruiz-Altaba for critical remarks and discussions on matters related to bootstrap symmetry breaking.

This work supported in part by the National Science Foundation: PHY 88-21039.

REFERENCES

1. Y. Nambu in *Rationale of Beings*, Festschrift in honor of G. Takeda (eds. K. Ishikawa et al., World Scientific, Singapore, 1986), p. 3., EFI preprint 85-56.

2. Y. Nambu in *New Theories in Physics* (Proc. XI Warsaw Symposium on Elementary Particle Physics, eds. Z. Adjuk et al., World Scientific, Singapore, 1989), p. 1.

   Y. Nambu, *New Trends in Strong Coupling Gauge Theories*, (Proc. 1988 International Workshop, Nagoya, ed. M. Bando et al., World Scientific, Singapore, 1989).

   Y. Nambu, EFI preprint 89-08 (unpublished).

   These papers contain some errors, as were kindly pointed out to me by Drs. M. Ruiz-Altaba and C. Hill.

3. Y. Nambu, to be published in the R. Dalitz Festschrift (EFI preprint 90-46).

4. V.A. Miransky, M. Tanabashi and K. Yamawaki, Mod. Phys. Lett., **A4** (1989) 1043; Phys. Lett., **B221** (1989) 177.

   W.A. Bardeen, C.T. Hill and M. Lindner, Phys. Rev., **D41** (1990) 1647.

THE EXCEPTIONAL SUPERSPACE AND THE QUADRATIC JORDAN FORMULATION OF QUANTUM MECHANICS*

M. Günaydin [†]
Dept. of Physics, Penn State University
University Park, PA 16802

Abstract

The Jordan superalgebra JF(6/4) is the unique exceptional Jordan superalgebra that has no realization in terms of Z_2 graded associative supermatrices. We study its symmetry supergroups considered as the basis of an exceptional superspace that is non-Clifford-algebraic. The superconformal group of this superspace is the exceptional supergroup F(4). The appropriate formalism for studying this non-associative superspace is the quadratic Jordan formulation of quantum mechanics. We discuss some of the implications of the exceptionality of JF(6/4).

---

*Work supported in part by the National Science Foundation under grant number PHY-8909549

[†]Bitnet: GXT@PSUVM

1 Introduction

Jordan algebras were introduced by Pascual Jordan in the early thirties with the hope of generalizing the formalism of quantum mechanics so as to be able to accommodate some of the newly observed phenomena such as beta-decay within that framework [1]. Subsequently, Jordan, von Neumann and Wigner showed that ,with but one possible exception, the Jordan formulation of quantum mechanics is equivalent to the formulation over an Hilbert space a la Dirac [2]. This exception is the Jordan algebra J_3^O of 3×3 hermitian octonionic matrices. About four and a half decades later Jordan superalgebras were defined and classified by Kac [3]. A decade after the work of Kac, A.S.Shtern proved that there exists only one finite dimensional exceptional Jordan superalgebra [4]. This exceptional Jordan superalgebra has no realization in terms of associative Z_2 graded matrices. The existence of an exceptional Jordan superalgebra raises many issues similar to the ones posed by the existence of an exceptional Jordan algebra. We shall address some of these issues in this article. Our main proposal is to consider the exceptional Jordan superalgebra as the basis of an exceptional superspace. We study its superrotation, super-Lorentz and superconformal symmetry algebras. Many of the methods developed for studying the exceptional Jordan algebra carry over to the study of the exceptional Jordan superalgebra. In particular, the quadratic Jordan formulation of quantum mechanics can be used to study Jordan superalgebras considered formally as Z_2 graded extensions of the usual Jordan formulation, which extends to the exceptional Jordan superalgebra just as it does to the exceptional Jordan algebra.

2 Space-time Supergroups and Jordan Superalgebras

The twistor formalism in four-dimensional space-time ($d = 4$) leads naturally to the representation of four vectors in terms of 2×2 Hermitian matrices over the field of complex numbers C. In particular, the coordinate four vectors can be represented as such. In this form the actions of four-dimensional space-time symmetry groups on the Minkowski space take on particularly elegant forms. For example, the action of the conformal group on the Minkowski coordinates can be realized as a group of linear fractional transformations of the corresponding 2×2 matrices [5]. The Hermitian matrices close under anti-commutation and form a Jordan algebra [1]. Then the rotation, Lorentz and conformal groups in $d = 4$ can be interpreted as the automorphism, reduced structure and Möbius groups of the Jordan algebra of 2×2 complex Hermitian matrices [6]. Conversely, this interpretation allows one to define the concepts of rotation, Lorentz and conformal groups for any Jordan algebra [6,7]. In the mathematics literature they have been studied under the names automorphism, reduced structure and superstructure groups ("super" in this case having nothing to do with supersymmetry) [8]. Denoting as J_n^A the Jordan algebra of $n \times n$ Hermitian matrices over the division algebra A, one finds the following symmetry groups associated with it:

J	RG	LG	CG
J_n^R	$SO(n)$	$SL(n,R)$	$Sp(2n,R)$
J_n^C	$SU(n)$	$SL(n,C)$	$SU(n,n)$
J_n^H	$USp(2n)$	$SU^*(2n)$	$SO^*(4n)$
J_3^O	F_4	$E_{6(-26)}$	$E_{7(-25)}$

The symbols R, C, H, O represent the four division algebras and RG, LG and CG are the rotation (automorphism), Lorentz (reduced structure), and conformal (Möbius) groups, respectively. In addition to the above Jordan algebras there is another infinite family of simple Jordan algebras [9], namely the Jordan algebra $\Gamma(d)$ of Dirac gamma matrices in d dimensions.

$$\{\Gamma_\mu, \Gamma_\nu\} = \delta_{\mu\nu} 1; \qquad \mu, \nu, \ldots = 1, 2, \ldots, d \qquad (1)$$

Their automorphism, reduced structure and Möbius groups are simply the rotation, Lorentz and conformal groups of $(d+1)$-dimensional Minkowski spacetimes. One finds the following special isomorphisms between the Jordan algebras of 2×2 Hermitian matrices over the four division algebras and the Jordan algebras of gamma matrices:

$$J_2^R \simeq \Gamma(2) \quad ; \quad J_2^C \simeq \Gamma(3) \quad ; \quad J_2^H \simeq \Gamma(5) \quad ; \quad J_2^O \simeq \Gamma(9) \qquad (2)$$

The Minkowski spacetimes they correspond to are precisely the critical dimensions for the existence of super Yang-Mills theories as well as of the classical Green-Schwarz superstrings. These Jordan algebras are all quadratic and their norm forms are precisely the quadratic invariants constructed using the Minkowski metric. The reduced structure group is defined as the invariance group of the norm form and the superstructure (conformal) group is simply the invariance group of the zero norm condition. Clearly, these definitions agree with the usual ones.

Jordan superalgebras are Z_2 graded algebras with a supersymmetric product and they have been classified by Kac [3]. One can similarly define their rotation, Lorentz and conformal supergroups as their automorphism, reduced structure and Möbius supergroups. A complete list of these supergroups was given in [6]. Below we list the simple Jordan superalgebras and their symmetry supergroups using a modified version of Kac's notation for labelling Jordan superalgebras. For example, the Jordan superalgebra of type X with m even elements and n odd elements is denoted as $JX(m/n)$.

JX	SRG	SLG	SCG
$JA(m^2 + n^2/2mn)$	$SU(m/n)$	$SU(m/n) \times SU(m/n)$	$SU(2m/2n)$
$JBC(\frac{1}{2}(m^2 + m) + (2n^2 - n)/2mn)$	$OSp(m/2n)$	$SU(m/2n)$	$OSp(4n/2m)$
$JD(m/2n)$	$OSp(m-1/2n)$	$OSp(m/2n)$	$OSp(m+2/2n)$
$JP(n^2/n^2)$	$P(n-1)$	$SU(n/n)$	$P(2n-1)$
$JQ(n^2/n^2)$	$Q(n-1) \times U(1)_F$	$Q(n-1) \times Q(n-1) \times U(1)_F$	$Q(2n-1)$
$JD(2/2)_\alpha$	$OSp(1/2)$	$SU(1/2)$	$D(2,1;\alpha)$
$JF(6/4)$	$OSp(1/2) \times OSp(1/2)$	$OSp(2/4)$	$F(4)$
$JK(1/2)$	$OSp(1/2)$	$SU(1/2)$	$SU(2/2)$

Above we denoted the super rotation, Lorentz and conformal groups of Jordan superalgebras as SRG, SLG and SCG, respectively. The term $U(1)_F$ denotes the "fermionic" $U(1)$ factor generated by a single odd generator [6,10].

Some of the supergroups appearing in the above list correspond to spacetime symmetry groups in various dimensions. In those cases one may consider the underlying Jordan superalgebra as the basis of a superspace by identifying the even subspace with the space-time coordinates and the odd subspace with the Grassmann coordinates. By multiplying the odd elements with anticommuting Grassmann coefficients one can work with the symmetric Jordan product in both sectors [6,10].

In the list of simple Jordan superalgebras one is truly unique, namely the exceptional Jordan superalgebra $JF(6/4)$. It is the only simple Jordan superalgebra which has no realization in terms of Z_2 graded associative supermatrices

[4]. Below we shall study $JF(6/4)$ considered as the basis of an exceptional superspace[11]. Its exceptionality implies that the corresponding supersymmetry is non-Clifford-algebraic. For a certain real form of $JF(6/4)$ the Möbius group $F(4)$ is simply the $N = 2$ superconformal group in five-dimensional Minkowski space (or the $N = 2$ anti-de Sitter supergroup in $d = 6$). It has the even subgroup $SO(5,2) \times SU(2)$. Another real form of $JF(6/4)$ leads to the real form of $F(4)$ which has the even subgroup $SO(7) \times SU(1,1)$. It corresponds to an exceptional $N = 8$ conformal supergroup in one dimension.

3   The Construction of Lie (Super)Algebras from Jordan (Super)Algebras

A Jordan algebra J is a non-associative algebra with a symmetric product $a \cdot b = b \cdot a$ satisfying the Jordan identity:

$$a \cdot (b \cdot a^2) = (a \cdot b) \cdot a^2, \quad \forall\ a, b \in J \tag{3}$$

A derivation D of J is an endomorphism of J such that

$$D(a \cdot b) = (Da) \cdot b + a \cdot (Db) \tag{4}$$

Derivations form a Lie algebra under commutation which is called the the derivation algebra $(DerJ)$ of J. If we denote the left and right multiplication by an element a as L_a and R_a, respectively, we have $L_a = R_a$ due to the symmetry of the Jordan product. The derivation condition can be rewritten in the form:

$$[\,D, L_a\,] = L_{Da} \tag{5}$$

By linearizing the Jordan identity one obtains the identity:

$$(a \cdot b) \cdot (c \cdot d) + (a \cdot c) \cdot (b \cdot d) + (a \cdot d) \cdot (b \cdot c) = a \cdot (b \cdot (c \cdot d)) + c \cdot (b \cdot (a \cdot d)) + d \cdot (b \cdot (a \cdot c)) \tag{6}$$

which is equivalent to the identity:

$$[L_{a \cdot b}, L_c] + [L_{c \cdot a}, L_b] + [L_{b \cdot c}, L_a] = 0 \tag{7}$$

and

$$[[L_a, L_b], L_c] = L_{[L_a, L_b]c} \tag{8}$$

It follows from the above identities that $[L_a, L_b]$ is a derivation of the Jordan algebra J. The derivations of a Jordan algebra can always be represented in this form [12], i.e they are all inner. Therefore we can formally write:

$$[L_J, L_J] = \text{Der } J \qquad (9)$$

Exponentiating the derivations one obtains the inner automorphisms of J

$$e^D(a \cdot b) = (e^D a) \cdot (e^D b) \qquad (10)$$

Thus the derivation algebra of J is simply the Lie algebra of its automorphism (rotation) group. The multiplication operators close under commutation into derivations of J. The Lie algebra generated by multiplications with elements of J and the derivations is called the structure algebra $St(J)$. The multiplication by the identity element of J commutes with all the other elements of $St(J)$. Factoring out this operator from $St(J)$ one is left with what is referred to as the reduced structure algebra $St(J)_0$ of J. It is simply the Lie algebra of the reduced structure group or what we called the Lorentz group (LG) of J in the previous section.

An alternative description of the structure algebra $St(J)$ is via the Jordan triple product which is defined as

$$\{abc\} \equiv a \cdot (b \cdot c) + (a \cdot b) \cdot c - b \cdot (a \cdot c) \qquad (11)$$

Using this triple product one defines a bilinear transformation S_{ab} on J as

$$S_{ab} c \equiv \{abc\}, \qquad \forall \ a, b, c \in J \qquad (12)$$

They generate $St(J)$ under commutation:

$$[S_{ab}, S_{cd}] = -S_{\{cda\}, b} + S_{a, \{dcb\}} \qquad (13)$$

This follows from the fact that S_{ab} is simply the sum of the derivation $D_{a,b}$ of J and the multiplication by $a \cdot b$:

$$S_{ab} = [L_a, L_b] + L_{a \cdot b} = D_{a,b} + L_{a \cdot b} \qquad (14)$$

The Tits-Koecher-Kantor (TKK) construction defines a 3-graded Lie algebra g over a given Jordan algebra J [13,14,15]:

$$g = g^{-1} \oplus g^0 \oplus g^{+1} \qquad (15)$$

The elements of grade $+1$ and grade -1 spaces are labelled by the elements of J (hence we are actually using two copies of J which are referred to as a Jordan pair [8]):

$$U_a \in g^{+1}$$
$$\tilde{U}_b \in g^{-1} \qquad \forall\ a,b \in J \qquad (16)$$
$$S_{ab} \in g^0$$

Then one defines the Lie product among these elements as

$$[U_a, \tilde{U}_b] \equiv S_{ab}$$
$$[S_{ab}, U_c] \equiv U_{\{abc\}}$$
$$[S_{ab}, \tilde{U}_c] \equiv -\tilde{U}_{\{bac\}} \qquad (17)$$
$$[S_{ab}, S_{cd}] \equiv S_{\{abc\},d} - S_{c,\{bad\}}$$

with all the other products vanishing. The Jacobi identities follow from the Jordan identity and the symmetry of the Jordan product. The resulting Lie algebra g is referred to as the superstructure algebra and is isomorphic to the Lie algebra of the superstructure (Möbius) group of J which we called the conformal group (CG) in the previous section. There is a non-linear realization of the conformal group CG on J as a group of linear fractional transformations [16].

The TKK construction has been generalized to the construction of Lie superalgebras from Jordan superalgebras [6,10]. A Jordan superalgebra is a Z_2 graded algebra $J = J^0 + J^1$ with a supercommutative product

$$a \cdot b = (-1)^{\alpha\beta} b \cdot a \qquad (18)$$
$$a \in J^\alpha, b \in J^\beta; \alpha, \beta = 0,1$$

which satisfies the identity

$$(-1)^{\alpha\gamma}[L_{a\cdot b}, L_c\} + (-1)^{\beta\alpha}[L_{b\cdot c}, L_a\} + (-1)^{\gamma\beta}[L_{c\cdot a}, L_b\} = 0 \qquad (19)$$

where the mixed bracket $[\,,\}$ denotes the usual Lie superbracket. The defining conditions imply the following identity:

$$[[L_a, L_b\}, L_c\} = (-1)^{\beta\gamma} L_{a\cdot(c\cdot b)-(a\cdot c)\cdot b} \qquad (20)$$

Multiplying the odd elements (grade one) with anticommuting Grassmann parameters and the even elements (grade zero) with ordinary complex parameters, the supercommutative product can be replaced with a commutative one and the TKK construction can be carried over to the super case in a straightforward manner [6,10].

4 The Exceptional Jordan Superalgebra $JF(6/4)$ and its Symmetry Supergroups

The construction outlined in the previous section can be inverted so as to construct Jordan algebras [15] and Jordan superalgebras [3] starting from Lie algebras and Lie superalgebras. In fact, the inverse TKK functor was used by Kac so as to define and classify Jordan superalgebras [3]. In [11] the inverse TKK functor was applied to the exceptional Lie superalgebra $F(4)$ in order to write down the multiplication table of the exceptional Jordan superalgebra $JF(6/4)$ in a basis that can be given a superspace interpretation. The exceptional Lie superalgebra $F(4)$ has been studied in [18,19,20] Referring to [11] for details we shall now give the multiplication table of $JF(6/4)$ and study its symmetries.

The even elements of $JF(6/4)$ belonging to grade zero subspace are denoted as S, B_0 and B_μ, ($\mu = 1, 2, 3, 4$). The odd elements belonging to the grade one subspace are denoted as Q_α, ($\alpha = 1, 2, 3, 4$). Their super-commutative Jordan products are:

$$B_\mu \cdot B_\nu = -\delta_{\mu\nu} B_0$$

$$B_0 \cdot B_\mu = B_\mu$$

$$B_0 \cdot B_0 = B_0$$

$$B_0 \cdot S = 0 = B_\mu \cdot S$$

$$S \cdot S = S \qquad (21)$$

$$Q_\alpha \cdot Q_\beta = (i\gamma_5\gamma_\mu C)_{\alpha\beta} B^\mu + (\gamma_5 C)_{\alpha\beta}(B_0 - 3S)$$

$$B_\mu \cdot Q_\alpha = \tfrac{i}{2}(\gamma_\mu)_{\alpha\beta} Q_\beta$$

$$B_0 \cdot Q_\alpha = \tfrac{1}{2} Q_\alpha$$

$$S \cdot Q_\alpha = \tfrac{1}{2} Q_\alpha$$

$$\mu, \nu, \ldots = 1, 2, 3, 4 \quad ; \quad \alpha, \beta, \ldots = 1, 2, 3, 4$$

The B_0 and S are the two idempotents and $I = B_0 + S$ is the identity element of $JF(6/4)$. The matrices γ_μ are the four-dimensional (Euclidean) Dirac gamma matrices and C is the charge conjugation matrix:

$$\{\gamma_\mu, \gamma_\nu\} = 2\delta_{\mu\nu}$$

$$\gamma_\mu C = -C \gamma_\mu^T \qquad (22)$$

In the above basis the elements of the derivation algebra $Der\ JF(6/4)$ are

$$D_{\mu,\nu} = [L_{B_\mu}, L_{B_\nu}] = -D_{\nu,\mu}$$

$$G_{\alpha,\beta} = \{L_{Q_\alpha}, L_{Q_\beta}\} = G_{\beta,\alpha}$$

$$F_{0,\alpha} = [L_{B_0}, L_{Q_\alpha}] \qquad (23)$$

$$F_{\mu,\alpha} = [L_{B_\mu}, L_{Q_\alpha}]$$

$$E_{S,\alpha} = [L_S, L_{Q_\alpha}]$$

They are however not all independent. We find that

$$F_{\mu,\alpha} = i(\gamma_\mu)_{\alpha\beta} F_{0,\beta}$$

$$E_{S,\alpha} = -F_{0,\alpha} \qquad (24)$$

$$G_{\alpha,\beta} = -\tfrac{1}{4}(\gamma_{\mu\nu}\gamma_5 C)_{\beta\alpha} D_{\mu,\nu}$$

They satisfy the supercommutation relations

$$[D_{\mu,\nu}, D_{\lambda,\rho}] = \delta_{\nu\lambda} D_{\mu,\rho} + \delta_{\mu\rho} D_{\nu,\lambda} - \delta_{\mu\lambda} D_{\nu,\rho} - \delta_{\nu,\rho} D_{\mu,\lambda}$$

$$[D_{\mu,\nu}, F_{0,\alpha}] = \tfrac{1}{2}(\gamma_{\mu\nu})_{\alpha\beta} F_{0,\beta} \qquad (25)$$

$$\{F_{0,\alpha}, F_{0,\beta}\} = -\tfrac{1}{8}(\gamma_5\gamma_{\mu\nu})_{\beta\alpha} D_{\mu\nu}$$

The even subalgebra is the Lie algebra of $SO(4)$ generated by $D_{\mu,\nu}$ and the full derivation algebra is the Lie superalgebra of $OSp(1/2) \times OSp(1/2)$.

$$\text{Der } JF(6/4) = OSp(1/2) \times OSp(1/2) \qquad (26)$$

The structure algebra of $JF(6/4)$ is generated by the derivations and the multiplications by the elements of $JF(6/4)$. They satisfy the following supercommutation relations

$$[D_{\mu,\nu}, L_{B_\lambda}] = \delta_{\mu\lambda} L_{B_\nu} - \delta_{\nu\lambda} L_{B_\mu}$$

$$[D_{\mu,\nu}, L_{Q_\alpha}] = \tfrac{1}{4}[\gamma_\mu, \gamma_\nu]_{\alpha\beta} L_{Q_\beta}$$

$$[F_{0,\alpha}, L_S] = \tfrac{1}{4} L_{Q_\alpha} \tag{27}$$

$$[F_{0,\alpha}, L_{B_0}] = -\tfrac{1}{4} L_{Q_\alpha}$$

$$[F_{0,\alpha}, L_{B_\mu}] = -\tfrac{i}{4}(\gamma_\mu)_{\alpha\beta} L_{Q_\beta}$$

$$[F_{0,\alpha}, L_{Q_\beta}] = \tfrac{1}{2}(i\gamma_5\gamma_\mu C)_{\alpha\beta} L_{B_\mu} + \tfrac{1}{2}(\gamma_5 C)_{\alpha\beta}(L_{B_0} + 3L_S)$$

plus those that define the derivations in terms of multiplication operators and those of $Der\ JF(6/4)$. The structure algebra with eleven even and eight odd generators is simply the Lie superalgebra of $OSp(2/4) \times U(1)$. The $U(1)$ generator is given by $L_I = L_{B_0} + L_S$. The even subgroup of the reduced structure group $OSp(2/4)$ is $SO(2) \times Sp(4)$.

At this point we should stress that different real forms of $JF(6/4)$ will have different real forms of the derivation and structure algebras. For example, there exist other real forms of $JF(6/4)$ for which the even subgroup $SO(4)$ of the automorphism group goes over to $SO(3,1)$ or to $SO(2,2)$. These different real forms of $JF(6/4)$ can be obtained via the application of the inverse TKK functor to different real forms of $F(4)$.

Finally, one can now rewrite $F(4)$ back as the Möbius (conformal) superalgebra of $JF(6/4)$ using the super generalization of the equations (17) as given in references [6,10]. For physical applications, more relevant is the non-linear realization of $F(4)$ as the supergroup of linear fractional transformations of $JF(6/4)$ in the sense of reference [16]. It simply corresponds to the superconformal group action on the corresponding superspace. Koecher's work on the linear fractional transformations [16] of Jordan algebras can be readily extended to Jordan superalgebras [6]. Koecher's realization and its super extension can be thought of as generalizations of the conformal group ($SO(4,2)$) action on the four dimensional Minkowski space-time. The Lie algebra of $SO(4,2)$ can be given a 3-graded decomposition

$$SO(4,2) = T_\mu \quad \oplus (J_{\mu\nu} + D) \quad \oplus K_\nu \tag{28}$$
$$= g^{-1} \quad \oplus \quad g^0 \quad \oplus g^{+1}$$

where grade -1 and grade $+1$ subspaces are spanned by generators of translations (T_μ) and special conformal transformations (K_ν), respectively. The grade zero subspace contains the generators of Lorentz transformations $(J_{\mu\nu})$ and dilatations (D). One can similarly decompose the conformal superalgebra g of a Jordan superalgebra J as

$$g = T_a \oplus S_{ab} \oplus K_b \tag{29}$$

where T_a, S_{ab} and K_b are the generators of translations, superstructure group and special superconformal tansformations of J, respectively. They act on a given element x of J as follows [16]:

$$T_a x = -a$$
$$S_{ab} x = \{abx\} \tag{30}$$
$$K_b x = \tfrac{1}{2}\{xbx\}$$

where $\{abx\}$ represents the (super)-Jordan triple product:

$$\{abx\} = a \cdot (b \cdot x) - (-1)^{\alpha\beta} b \cdot (a \cdot x) + (a \cdot b) \cdot x \tag{31}$$

They satisfy the super-commutation relations:

$$[T_a, K_b\} = S_{ab}$$
$$[S_{ab}, T_c\} = T_{\{abc\}}$$
$$[K_c, S_{ab}\} = K_{\{bac\}} \tag{32}$$
$$[T_a, T_b\} = 0 = [K_a, K_b\}$$

Clearly, T_a and K_a are odd or even generators depending on whether or not a is an odd or an even element of J, respectively. The generator S_{ab} is odd if either a or b is odd and is even otherwise. By substituting the basis elements of $JF(6/4)$ in the above formulas we get the realization of the exceptional superalgebra $F(4)$ on $JF(6/4)$ as its superconformal algebra. We should also note that one can rewrite the above realization of the super-Mobius algebra of a Jordan superalgebra in terms of differential operators as follows. One chooses a basis (e_i, o_α) for the Jordan superalgebra J and expands an element x of J as $x = x^i e_i + \theta^\alpha o_\alpha$ where θ^α are Grassmann parameters and x^i are real parameters. Then the action of the generators T_a, K_a and S_{ab} can be represented by differential operators of the form:

$$T_a \longleftrightarrow (\partial_i, \partial_\alpha)$$

$$S_{ab} \longleftrightarrow (p_1^i \partial_i, q_1^\alpha \partial_\alpha) \tag{33}$$

$$K_a \longleftrightarrow (p_2^i \partial_i, q_2^\alpha \partial_\alpha)$$

where $p_1^i(q_1^\alpha)$ and $p_2^i(q_2^\alpha)$ are even(odd) polynomials of degree one and two, respectively.

5 The Quadratic Jordan Formulation of Quantum Mechanics and The Exceptional Jordan Superalgebra

A Jordan algebra is said to be special if it can be represented in terms of associative matrices with the Jordan product defined as one half the anticommutator. Otherwise, it is said to be exceptional. In their classic work, Jordan, Von Neumann and Wigner proved that with but one possible exception all Jordan algebras are special [2]. This possible exception is the Jordan algebra J_3^O of 3×3 Hermitian octonionic matrices, whose exceptionality was proven by Albert [21]. In the Eighties Zelmanov extended these results to the infinite dimensional case and proved that there are no infinite dimensional exceptional Jordan algebras [22].

In the Jordan formulation of quantum mechanics the observables and the density matrices representing a physical system are elements of a Jordan algebra. If the underlying Jordan algebra is special, then the Jordan formulation is equivalent to the Dirac formulation of quantum mechanics over a Hilbert space. The existence of an exceptional Jordan algebra raised the question as to whether or not one can formulate quantum mechanics over the exceptional Jordan algebra satisfying all the axioms of Von Neumann. This was answered in the affirmative in the early seventies [23]. This so-called octonionic quantum mechanics has no Hilbert space formulation. The axioms of quantum mechanics as formulated by Von Neumann are equivalent to the axioms of projective geometry [24]. The projective geometry corresponding to the octonionic quantum mechanics is non-desarguian [23], which implies that it cannot be embedded in a higher dimensional projective geometry. This is consistent with the exceptionality of J_3^O.

The Jordan formulation was subsequently extended to the quadratic Jordan formulation of quantum mechanics which is applicable to the octonionic case as well [6]. The starting point of this formulation is the simple observation that the action of an hermitian operator H on a state $|\alpha>$ in a Hilbert space should be represented by the quadratic operator Π_H in the Jordan formulation i.e

$$H: |\alpha> \longrightarrow H|\alpha> \Longrightarrow H|\alpha><\alpha|H = HP_\alpha H = \{HP_\alpha H\} = \Pi_H P_\alpha \tag{34}$$

where $\{abc\}$ denotes the Jordan triple product and P_α is the projection operator onto the state $|\alpha>$:

$$P_\alpha = |\alpha><\alpha| \quad ; \quad P_\alpha^2 = P_\alpha \tag{35}$$

For special Jordan algebras we have $\{aba\} = aba$ where aba stands for the associative product of the matrices (or the operators) a, b and a. Recall also that in the original Jordan formulation the action of an observable H is simply given by $H \cdot P_\alpha$ and the expectation value of H in the state $|\alpha>$ is

$$<\alpha|H|\alpha> = Tr\, P_\alpha \cdot H \tag{36}$$

where Tr stands for the trace form. In the quadratic Jordan formulation the expectation value is given by:

$$<\alpha|H|\alpha> = Tr\, \Pi_{P_\alpha} H \tag{37}$$

In the case of a density matrix $\rho = \sum_i \alpha_i P_i$ $(Tr\, \rho = 1)$ corresponding to an incoherent superposition of pure states P_i $(P_i^2 = P_i, Tr\, P_i = 1)$ the expectation value of H is then

$$<H>_\rho = Tr\, \rho \cdot H = Tr \sum_i \alpha_i \Pi_{P_i} H \tag{38}$$

All the measurable quantities such as expectation values and transition probabilities can be expressed in terms of the Jordan triple product and the trace function. For example, the probability function $\Pi_{\alpha\beta}$ of the overlap of two states $|\alpha>$ and $|\beta>$ is given by

$$\begin{aligned}\Pi_{\alpha\beta} &= |<\alpha|\beta>|^2 = Tr\, P_\alpha \cdot P_\beta = Tr\, \{P_\alpha P_\beta P_\alpha\} \\ &= Tr\, \{P_\beta P_\alpha P_\beta\} = Tr\, \Pi_{P_\alpha} P_\beta = Tr\, \Pi_{P_\beta} P_\alpha\end{aligned} \tag{39}$$

The transition probability $|<\alpha|H|\beta>|^2$ is expressed as:

$$\begin{aligned}|<\alpha|H|\beta>|^2 &= <\alpha|H|\beta><\beta|H|\alpha> \\ &= Tr\, |\alpha><\alpha|H|\beta><\beta|H &&= Tr\, P_\alpha \cdot (\Pi_H P_\beta) \\ &= Tr\, H|\alpha><\alpha|H|\beta><\beta| &&= Tr\, (\Pi_H P_\alpha) \cdot P_\beta \\ &= Tr\, \Pi_{P_\alpha} \Pi_H P_\beta &&= Tr\, \Pi_{P_\beta} \Pi_H P_\alpha \\ &= Tr\, \{P_\alpha\{HP_\beta H\}P_\alpha\} &&= Tr\, \{P_\beta\{HP_\alpha H\}P_\beta\}\end{aligned}$$
(40)

Note that the quadratic operator Π is hermitian with respect to the trace form:

$$(A, \Pi_H B) = Tr\, A \cdot \Pi_H B = (\Pi_H A, B) = Tr\, (\Pi_H A) \cdot B \tag{41}$$

Thus it is also hermitian with respect to the transition probability. The non-measurable transition matrix element $<\alpha|H|\beta>$ can not be expressed in terms of the Jordan product and the trace form. In the original Jordan formulation the commutator of two observables H_1 and H_2 can not be expressed in terms of the Jordan product since $[H_1, H_2]$ is a non-hermitian operator. To define the compatibility of two observables H_1 and H_2 one introduces their associator with an arbitrary hermitian operator H

$$[H_1, H, H_2] \equiv (H_1 \cdot H) \cdot H_2 - H_1 \cdot (H \cdot H_2) \qquad (42)$$

which reduces to

$$\frac{1}{4}[H, [H_1, H_2]] \qquad (43)$$

for special Jordan algebras with the Jordan product $A \cdot B = \frac{1}{2}(AB + BA)$. Then the two observables are said to be compatible if their associator with H vanishes for all H. On the other hand the commutator of two observables can be defined in the quadratic formulation. Even though $[H_1, H_2]$ is not hermitian one can define a quadratic action corresponding to it that is hermitian. Assume first that we are dealing with hermitian operators acting on a complex Hilbert space. The quadratic action corresponding to the action of $[H_1, H_2]$ on a state $|\alpha>$ is simply

$$[H_1, H_2] \, |\alpha> \longrightarrow [H_1, H_2] \, |\alpha><\alpha| \, [H_1, H_2]^\dagger \qquad (44)$$

It can be rewritten in terms of the Jordan triple product as follows:

$$[H_1, H_2] \, |\alpha><\alpha| \, [H_1, H_2]^\dagger$$

$$= [H_1, H_2] P_\alpha [H_2, H_1]$$

$$= (H_1 H_2 P_\alpha H_2 H_1 + H_2 H_1 P_\alpha H_1 H_2 - H_1 H_2 P_\alpha H_1 H_2 - H_2 H_1 P_\alpha H_2 H_1) \qquad (45)$$

$$= (2\Pi_{H_1} \Pi_{H_2} + 2\Pi_{H_2} \Pi_{H_1} - 4\Pi_{H_1 \cdot H_2}) P_\alpha$$

$$= 4(\Pi_{H_1} \cdot \Pi_{H_2} - \Pi_{H_1 \cdot H_2}) P_\alpha$$

Therefore the quadratic operator corresponding to the commutator of two observables H_1 and H_2 is

$$\Pi_{[H_1, H_2]} \equiv 4(\Pi_{H_1} \cdot \Pi_{H_2} - \Pi_{H_1 \cdot H_2}) \qquad (46)$$

Having obtained this result for some special Jordan algebras in terms of the triple product we can now use the same definition for all Jordan algebras including the exceptional one. Thus in the quadratic Jordan formulation the two observables H_1 and H_2 are said to be compatible when

$$\Pi_{H_1} \cdot \Pi_{H_2} = \Pi_{H_1 \cdot H_2} \qquad (47)$$

To summarize let us list the formulas for physically important quantities in the different formulations of quantum mechanics:

HilbertSpace Formulation	Jordan Formulation	QuadraticJordan Formulation
$\| \alpha >$	$\| \alpha > < \alpha \| = P_\alpha$	P_α
$H \| \alpha >$	$H \cdot P_\alpha$	$\Pi_H P_\alpha = \{H P_\alpha H\}$
$< \alpha \| H \| \beta >$	?	?
$< \alpha \| H \| \alpha >$	$Tr\, H \cdot P_\alpha$	$Tr\, \Pi_{P_\alpha} H = Tr\, \{P_\alpha H P_\alpha\}$
$\|< \alpha \| H \| \beta >\|^2$	$Tr\, P_\alpha \cdot \{H P_\beta H\}$	$Tr\, \Pi_{P_\alpha} \Pi_H P_\beta$
	$= Tr\, \{H P_\alpha H\} \cdot P_\beta$	$= Tr\, \Pi_{P_\beta} \Pi_H P_\alpha$
$[H_1, H_2]$	?	$4(\Pi_{H_1} \cdot \Pi_{H_2} - \Pi_{H_1 \cdot H_2})$

The importance of the quadratic formulation of Jordan algebras lies in the fact that it is applicable to Jordan algebras over fields of any characteristic [8]. Therefore one can formally extend the quadratic formulation of quantum mechanics over arbitrary fields as well as to Jordan superalgebras. In the latter case the odd elements of the Jordan superalgebra are to be taken with Grassmann coefficients. However, one then loses the probability interpretation of quantum mechanics since the the trace function is no longer a real number. The quadratic Jordan formulation extends to the exceptional Jordan superalgebra just as it does to the exceptional Jordan algebra. One possible way out of the problem of having a probability function which is an hypercomplex number is to identify the odd subspace with unobservable (spinorial) degrees of freedom of a physical system. It is an open problem whether this interpretation leads to a physically acceptable extension of quantum mechanics by "odd" degrees of freedom.

6 On the Exceptionality of $JF(6/4)$ as the Basis of a Superspace

In the Eighties the exceptional Jordan algebra J_3^O has made its appearance within the framework of supergravity theories through the work of Günaydin, Sierra and Townsend (GST) [25]. In their construction and classification of $N = 2$ Maxwell-Einstein supergravity theories, GST showed that there exist four remarkable theories of this type that are uniquely determined by simple Jordan algebras of degree 3. These are the Jordan algebras J_3^R, J_3^C, J_3^H and the exceptional Jordan algebra J_3^O. Their symmetry groups in five, four and three space-time dimensions give the famous Magic Square of Freudenthal, Rozenfeld and Tits [26]. The exceptional theory defined by J_3^O shares all the remarkable features of the maximally extended N = 8 supergravity theory in the respective dimensions. More recently, several authors have speculated on the possible role J_3^O may play in the framework of string theories [27].

In going over to the Jordan superalgebras most of the concepts and definitions for ordinary Jordan algebras carry over in a natural way. For example, consider a Z_2 graded associative algebra $R = R^0 + R^1$. The elements of R define a Jordan superalgebra under the superanticommutator:

$$A \cdot B \equiv \frac{1}{2}(AB + (-1)^{\alpha\beta} BA); \quad \forall\ A, B \in R \qquad (48)$$

In most applications the algebra R is the algebra of operators acting on a Z_2-graded vector space $V = V^0 + V^1$. Hence one can represent them as supermatrices [28]. For space-time supersymmetry the corresponding vector space V is referred to as the superspace.

A Jordan superalgebra is then special if it can be represented in terms of associative Z_2 graded supermatrices with the product being one half the superanticommutator. All simple Jordan superalgebras are special except for $JF(6/4)$ [4]. (Here we are only considering Jordan superalgebras whose odd subspaces J^1 are not empty.) The existence of an exceptional Jordan superalgebra raises many interesting questions. For example if we take it to be the basis of an exceptional superspace then this superspace can not have any realization as an ordinary Z_2 graded vector space on which the supersymmetry generators act linearly. For special Jordan superalgebras this can always be done. In the language of ordinary quantum mechanics this means that for special Jordan algebras one may choose to work either with density matrices or with vectors in a Hilbert space and these two formulations are equivalent. One major advantage of the Hilbert or vector space formulation is the fact that one can define tensor products of vector spaces unambiguously. For ordinary superspace this means that one can develop a tensor calculus. On the other hand, defining tensor products of Jordan algebras is a notoriously difficult problem; e.g, the tensor product of two Jordan algebras is in general not a Jordan algebra. Therefore, if we were to consider $JF(6/4)$ as the basis of an exceptional superspace then it would be necessary to formulate it in the language of ordinary Z_2 graded vector spaces in order to develop a tensor calculus over it. Assume that there exists such a "generalized vector space" and let us refer to it as the exceptional supermodule. The action of the supersymmetry operators as well as the tensor product operation over it will then have to be non-associative. In other words, the tensor calculus corresponding to this exceptional superspace would have to be non-associative. Furthermore, the exceptional $N = 2$ supergravity theories discovered by GST [25] cannot be written down in the language of "ordinary" conformal tensor calculus [29]. Interestingly enough, the exceptional $N = 2$ theories originate in five space-time dimensions, and the real form of $F(4)$ with the even subgroup $SO(5,2) \times SU(2)$ is simply the $N = 2$ conformal superalgebra in that dimension. This suggests strongly that the underlying superconformal tensor calculus of these theories may be the "non-associative" tensor calculus associated with $JF(6/4)$. One way to approach this problem is to look at the unitary representations of $F(4)$ and try to interpret them in the language of a ten-dimensional superspace corresponding to $JF(6/4)$. We should note that a pure gauged anti-de Sitter supergravity in $d = 6$ with $F(4)$ symmetry was constructed in [30]. Matter couplings of that theory as well as the conformal supergravity in $d = 5$ with $F(4)$ symmetry are yet to be constructed.

Finally, the algebra $JF(6/4)$ considered as the basis of a superspace corresponds to six bosonic and four "fermionic" (Grassmann) coordinates. Because of its potential relevance to the superstring one may want to know if there exists a larger algebraic structure, with ten bosonic elements and sixteen fermionic ones, that can be considered as the basis of a superspace. From Kac's classification we know that this algebraic structure can not be a simple Jordan superalgebra, though it may contain $JF(6/4)$ as a subalgebra. If such an algebraic structure exists then it may give us the clue as to how to generalize conformal supersymmetry to ten space-time dimensions.

References

[1] P. Jordan, Nach. Ges. Wiss. Göttingen (1932), 209 ; Z. Physik *80* (1933), 285.

[2] P. Jordan, J. von Neumann and E. Wigner, Ann. Math. *36* (1934), 29.

[3] V. Kac, Comm. in Algebra *5* (1977), 1375.

[4] A.S. Shtern, Funkt. Analiz. i Ego Prilozheniya *21* (1987), 93.

[5] F. Gürsey, Nuovo Cimento *3* (1956), 988.

[6] M. Günaydin, Ann. of Israel Physical Soc. *3* (1980), 279.

[7] M. Günaydin, Nuovo Cimento *29A* (1975), 467.

[8] See the excellent review by K. McCrimmon, Bull. Am. Math. Soc. *84* (1978), 612 and the references therein.

[9] R.D. Schafer, " An Introduction to Non-associative Algebras ", Academic Press, New York 1966.

[10] I. Bars and M. Günaydin, J. Math. Phys. *20* (1979), 1977.

[11] M. Günaydin, Jour. Math. Phys. *31* (1990) 1776.

[12] N. Jacobson, "Structure and Representations of Jordan Algebras", Amer. Math. Soc. Coll. Pub. Vol.39 (Rhode Island, 1968).

[13] J. Tits, Nederl. Akad. van Wetens. *65* (1962), 530.

[14] M. Koecher, Amer. J. Math. *89* (1967) 787.

[15] I.L. Kantor, Sov. Math. Dok. *5* (1964), 1404.

[16] M. Koecher, Inv. Math. *3* (1967), 136.

[17] M. Günaydin, J. Math. Phys. *30* (1989), 937.

[18] W. Nahm, V. Rittenberg and M. Scheunert, J. Math. Phys. *9* (1976), 1640.

[19] B.S. De Witt and P. van Nieuwenhuizen, J. Math. Phys. *23* (1982), 1953.

[20] A. Sudbery, Jour. Math. Phys. *24* (1983), 1986.

[21] A.A. Albert, Ann. Math. *36* (1934), 65.

[22] E.I. Zelmanov, Sib. Math. Jour. *24* (1983), 89.

[23] M. Günaydin, C. Piron and H. Ruegg, Comm. Math. Phys. *61* (1978), 69.

[24] C. Piron, " Foundations of Quantum Physics ", Benjamin Press, (Mass. 1976).

[25] M. Günaydin, G. Sierra and P.K. Townsend, Phys. Lett. *133B*(1983), 72; Nucl. Phys. *B242* (1984), 244; ibid *B253* (1985), 573.

[26] H. Freudenthal, Nederl. Akad. Wetensch. Proc. *A62* (1959), 447; B.A. Rozenfeld, Dokl. Akad. Nauk. SSSR *106* (1956),600; J. Tits, Mem. Acad. Roy. Belg. Sci. *29* (1955) fasc.3.

[27] D.B. Fairlie and C. Manouge, Phys. Rev. *D34* (1986), 1832; G. Sierra, Class. Quant. Grav. *4* (1987), 385; P. Goddard, W. Nahm, D. Olive, H. Ruegg and A. Schwimmer, Comm. Math. Phys. *112* (1987), 385; F. Gürsey, Mod. Phys. Lett. *A2* (1987), 967; G. Chapline and M. Günaydin, UCRL Preprint 95290 (1986), unpublished; E. Corrigan and T.J. Hollowood, Phys. Lett.*B203* (1988),47; M. Günaydin and S.J. Hyun, Phys. Lett. *B209* (1988), 498.

[28] M. Scheunert, " The Theory of Lie Superalgebras ", Lecture Notes in Math. *716*, Springer Verlag, New York, 1979.

[29] B. de Wit and A. van Proeyen, private communication.

[30] L.J. Romans, Nucl. Phys. *B269* (1986), 691.

ALGEBRA OF REPARAMETRIZATION-INVARIANT AND NORMAL ORDERED OPERATORS IN OPEN STRING FIELD THEORY

P. Ramond*

Department of Physics, University of Florida

Gainesville, Florida 32611

Dedicated to Murray Gell-Mann on the occasion of his sixtieth birthday

ABSTRACT

We study the algebra of normal ordered and reparametrization invariant operators of the open bosonic string field theory. These, besides the Poincaré group generators, include the ghost number operator and two translationally invariant symmetric second-rank space-time tensors. The BRST operator of string field theory is the trace of the fermionic one, and the second is the BRST transform of the former. Their algebra closes only when certain Lorentz non-invariant projections over the fermionic tensor are taken. There are many inequivalent such algebras, corresponding to manifest Lorentz invariance in lower dimensions. Some of these contain, besides the Lorentz invariant BRST, another nilpotent operator. We provide an example where Lorentz invariance is manifest in $1+1$ dimensions.

---

* Supported in part by the United States Department of Energy under grant DE-FG05-86-ER40272

I recall that the first time Murray asked me to come and talk at Caltech sixteen years ago, the subject was one we both shared an interest in: covariant string field theory. After all these years, things have not changed so much as to force me to change the subject matter: *Plus ça change, plus c'est la même chose.* I recall Murray inquiring at the time as to the underlying principle behind string theory. His penetrating question is still valid and unanswered today.

In a series of publications[1], we set about to list normal-ordered operators which are invariant under complex reparametrizations. In open string field theory, these include familiar operators such as the ghost number operator, the Poincaré generators, and the BRST charge. Interestingly, several unexpected operators appeared as well: two symmetric space-time tensors, one bosonic, one fermionic, with the trace of the fermionic tensor being none other than the BRST operator. In the case of the open superstring the same list was reproduced, with the addition of two space-time vectors when fermionized superconformal ghosts are used.

In building these operators, we have been guided by our pre-BRST program[2], where dynamics is to be formulated in a reparametrization invariant way. On the other hand, it is by now well known that the Open String Field Theory of Witten[3] is predicated not on reparametrization invariance, but on BRST invariance[4]. In fact, reparametrization invariance is explicitly broken in the interaction vertex. However there are several indications this is an incomplete picture as Witten's particular generalization of BRST does not lead to unitary answers for the open string without taking account of closed string poles, and all efforts to apply the BRST idea to closed strings have so far proven fallow.

It seems to us that the fundamental principle upon which string theory is built has not yet been formulated. In fact, it is easy to believe that there is no Lagrangian for closed string field theory. Whatever that principle may be, it would be safe to assume that it includes reparametrization invariance. Proposals for such a principle have been advanced, but none have proven themselves against the acid test of closed string field theory. We believe that, in this state of conceptual confusion, one should study all the algebraic structures which appear in the theory, hoping to recognize in their edifice the seeds of the real theory. While we recognize that in the Siegel-Witten approach, BRST invariance should suffice to explain quantum behavior, we are of the opinion that reparametrization invariance is more fundamental, and more closely aligned to the hitherto unknown invariance principle from which this whole structure emerges. Thus, we are intrigued by the appearance of operators which are more general than BRST. While we do not understand their role, we find it remarkable that they satisfy a closed superalgebra, which we proceed to discuss.

Operators in the Open String Field Theory

Open string functionals are acted upon by a set of kinematical operators (M) which reparametrize the real parameter σ along the string. These operators form a closed, anomaly-free algebra (it is identified with a subalgebra of the Virasoro algebra). Dynamical operators (K), *i.e.* those containing time-derivatives, can be formed which not only transform covariantly under reparametrizations up to an anomaly, but also commute *modulo* reparametrizations. Except for the anomaly, the algebraic structure is akin to that of the Lorentz algebra in four dimensions, which has a kinematical (*i.e.* no time derivatives) SU(2) rotation subgroup, and dynamical boost operators which transform covariantly under, and commute up to rotations. The analogy could even be carried further: in Minkowski space, rotation and boosts can be combined to build two SU(2) algebras which can be related to one another by conjugation or parity. In the open string case, the two algebras M and K can similarly be combined into two algebras which are related by parity in σ the parameter along the string, so that these two algebras are in fact the same, the open string Visasoro algebra. The analogy holds when one regards the transition from the three dimensional rotation to the Lorentz group in four dimensions as going from a rotation with real angles to one with complex angles (certainly true in the spinor representation), while in the open string case, the transition from M to K results in going from real to complex reparametrizations.

In Euclidean space, however, rotation and "boosts" can also be combined, to yield two completely *independent* algebras. This case is analogous to that of the closed string where the combination of M and K form two commuting algebras which are totally independent – these are the left and right movers. However the mystery of the closed string is deeper because the separation of left and right movers does not allow for rigid rotations, resulting in a theory where these two sectors are related. Here the situation reminds one of chiral spontaneous symmetry breaking where hitherto independent left and right sectors are mixed by the order parameter which then serves as a mass. Thus our approach, by giving a fundamental role to the reparametrizations, suggests that we view the chirality violating condition $L_0 - \bar{L}_0 = 0$ as a symmetry breaking condition imposed on the chiral reparametrization group. We are then faced with two choices, the uninteresting one views it as an explicit symmetry breaking term, the second views it as spontaneous symmetry breaking. The latter point of view would then imply that *the* full gauge invariant closed string field theory must include extra degrees of freedom to describe the order parameter, as in Higgs theory. (An algebraic analogy could be made with the hydrogen atom in a strong external magnetic field which breaks the $SO(3) \times SO(3)$ symmetry, and forces the system in a definite magnetic state). We note that under the rigid rotation generated by $L_0 - \bar{L}_0$, the string parameter σ simply gets translated by a constant, just like a Nambu-Goldstone boson would! In this picture, the vacuum state is characterized by the left-right breaking condition, emphasizing[5] its role in the formulation of a gauge invariant closed-string field theory; this is in accordance with some recent speculations based on topological field theories[6]. In the following we restrict our discussion to open strings, and hope to return to these issues for the closed string in the near future.

Of course the program we have just outlined does not make any sense unless the algebras are anomaly free. The reparametrization algebra for open strings which includes both dynamical and kinematical parts is anomalous unless it is extended to include ghosts – in this case the anomaly cancels with twenty-six coordinates, as is well known. The algebra is realized on the 26 coodinates $x^\mu(\sigma)$ and bosonized ghost $\phi(\sigma)$. The kinematical algebra is generated by

$$M_f = -i \int_0^\pi \frac{d\sigma}{\pi} [f(x'^\mu \frac{\delta}{\delta x^\mu} + \phi' \frac{\delta}{\delta \phi}) + wf' \frac{\delta}{\delta \phi}],$$

where w is the weight associated with ϕ and the real function $f(\sigma)$ is proportional to the infinitesimal reparametrization; the prime denotes differentiation with respect to σ. These generators form the algebra

$$[M_f, M_g] = iM_{f'g-fg'},$$

with the function f vanishing at the end points $\sigma = 0, \pi$.

The same algebra can be realized on more general functions: introduce the combinations

$$x'^\mu_{L,R} = x'^\mu \pm i\frac{\delta}{\delta x_\mu} \; ; \; \phi'_{L,R} = \phi' \mp i\frac{\delta}{\delta \phi},$$

in terms of which the operators

$$M^L_f = \int_{-\pi}^{+\pi} d\sigma f(\sigma)[x'_L \cdot x'_L - \phi'_L \phi'_L - 3\phi''_L]$$

form an algebra isomorphic to the kinematical one, except that it generates complex reparametrizations. One can also construct M^R_f which commutes with M^L_f. In the open string theory, it corresponds to a parity rearrangement and does not generate any new algebra.

Our intent is to study operators which are invariant under this algebra, as well as normal-ordered. Our previous list contains

$$\text{Momentum}: P_\mu = \int_0^\pi \frac{d\sigma}{\pi} \frac{\delta}{\delta x^\mu(\sigma)},$$

$$\text{Ghost momentum (ghost number)}: N = \int_0^\pi \frac{d\sigma}{\pi} \frac{\delta}{\delta \phi(\sigma)},$$

$$\text{Lorentz generators}: M_{\mu\nu} = \int_0^\pi \frac{d\sigma}{\pi} (x_\mu \frac{\delta}{\delta x^\nu} - x_\nu \frac{\delta}{\delta x^\mu}),$$

$$\text{Symmetric tensor}: Q^{\mu\nu} = \int_{-\pi}^\pi \frac{d\sigma}{2\pi} : e^{\phi_L} [x_L'^\mu x_L'^\nu - \frac{g^{\mu\nu}}{26}(\phi_L'^2 + 3\phi_L'')] : .$$

We augment this list by a symmetric traceless bosonic tensor

$$B^{\mu\nu} = \int_{-\pi}^{+\pi} \frac{d\sigma}{2\pi} : e^{2\phi_L}(x_L'^\mu x_L'^\nu - \frac{g^{\mu\nu}}{26} x' \cdot x') : .$$

All of these operators are normal ordered and invariant under the complex reparametrization algebra. We proceed to study the algebra obeyed by these special quantities.

Clearly, P_μ and $M_{\mu\nu}$ form the algebra of the Poincaré group, and commute with ghost number, while $Q^{\mu\nu}$ has ghost number 1 and $B^{\mu\nu}$ ghost number 2:

$$[N, Q^{\mu\nu}] = Q^{\mu\nu} \; ; \; [N, B^{\mu\nu}] = 2B^{\mu\nu}.$$

Furthermore, the translationally invariant $Q^{\mu\nu}$ and $B^{\mu\nu}$ transform as second rank tensors under the Lorentz group.

On the other hand, the $Q^{\mu\nu}$ obey a more complicated anticommuting algebra

$$\{Q^{\mu\nu}, Q^{\rho\sigma}\} = -i(g^{\nu\rho}B^{\mu\sigma} + g^{\mu\rho}B^{\nu\sigma} + g^{\mu\sigma}B^{\nu\rho} + g^{\nu\sigma}B^{\mu\rho})$$
$$+ \frac{2}{13}[(g^{\mu\nu}B^{\rho\sigma} + g^{\rho\sigma}B^{\mu\nu}) - (g^{\mu\sigma}g^{\nu\rho} + g^{\mu\rho}g^{\nu\sigma} - \frac{1}{13}g^{\mu\nu}g^{\rho\sigma})C],$$

where
$$C = \int_{-\pi}^{+\pi} \frac{d\sigma}{2\pi} e^{2\phi_L} x'_L \cdot x'_L \ ;$$

it is an invariant tensor, but it is *not* normal ordered. Hence the algebra of the $Q^{\mu\nu}$ does not close on normal-ordered invariant operators, which is hardly surprising. The algebra is completed by noting that $B^{\mu\nu}$ commutes with itself and with $Q^{\mu\nu}$

$$[B^{\mu\nu}, Q^{\rho\sigma}] = [B^{\mu\nu}, B^{\rho\sigma}] = 0.$$

In order to determine the largest non-anomalous algebra among our operators, let us introduce a set of 26×26 matrices $\alpha^I{}_{\mu\nu}$ such that the projections

$$Q^I \equiv \alpha^I{}_{\mu\nu} Q^{\mu\nu} \equiv \text{Tr}(\alpha^I \mathbf{Q})$$

satisfy a non-anomalous algebra. It is easy to see that

$$\{Q^I, Q^J\} = 2i\{\alpha^I, \alpha^J\}_{\mu\nu} B^{\mu\nu} + \frac{i}{13}\{B^J \text{Tr}(\alpha^I) + B^I \text{Tr}(\alpha^J)\}$$
$$- \frac{i}{13}\{\text{Tr}(\alpha^I \alpha^J) - \frac{1}{26} \text{Tr}(\alpha^I)\text{Tr}(\alpha^J)\} C,$$

where $B^I = \alpha^I{}_{\mu\nu} B^{\mu\nu}$. Thus the algebra will not be anomalous as long as we require

$$\text{Tr}(\alpha^I \alpha^J) = \frac{1}{26} \text{Tr}(\alpha^I) \text{Tr}(\alpha^J) ,$$

where the trace operation is meant to act between covariant and contravariant indices, *i.e.*

$$\text{Tr } \alpha^I = (\alpha^I)^\mu{}_\mu .$$

First we note that this condition is obeyed by the metric itself

$$\text{Tr}(\mathbf{g}\,\mathbf{g}) = \frac{1}{26} \text{Tr}\mathbf{g}\,\text{Tr}\mathbf{g} .$$

In fact the only Lorentz-invariant projection

$$Q = g_{\mu\nu} Q^{\mu\nu}$$

is the nilpotent BRST charge

$$\{Q, Q\} = 0 .$$

Similarly, we see that $B^{\mu\nu}$ is the BRST transform of $Q^{\mu\nu}$

$$\{Q, Q^{\mu\nu}\} = 2i B^{\mu\nu} ,$$

so that it commutes with Q.

It is enough to concentrate on the set of 26×26 matrices which satisfy the following conditions

$$(\alpha^I)_{\mu\nu} = (\alpha^I)_{\nu\mu} \ ; \ (\alpha^I)^\mu{}_\mu = 0 \ ;$$

$$(\alpha^I)_\mu{}^\rho (\alpha^J)_\rho{}^\mu = 0 .$$

These matrices do not exist in Euclidean space. K, the number of matrices obeying these properties in D dimensions, can be written as the number of symmetric traceless matrices less the number of constraints, $i.e.$

$$K = \frac{D(D+1)}{2} - 1 - \frac{K(K+1)}{2} ,$$

which is satisfied when $K = D - 1$. Thus, including the Lorentz invariant BRST charge, there are at most 26 such Q^I which satisfy a non-anomalous algebra. Of course none of these, except the trace, commute with the full Lorentz group.

The traceless projections obey the algebra

$$\{Q^I, Q^J\} = 2i\{\alpha^I, \alpha^J\}_{\mu\nu} B^{\mu\nu} .$$

However, these projections can be transformed by Lorentz transformations into other operators which are not in the set $\{Q^I\}$ since the Lorentz algebra has been broken by the very choice of the α^I matrices. Neglecting the trace, let us expand the 350 symmetric matrices $Q_{\mu\nu}$ in the convenient basis

$$Q_{\mu\nu} = \sum_{I=1}^{25} \beta^I{}_{\mu\nu} Q^I + \sum_{I=1}^{25} \alpha^I{}_{\mu\nu} R^I + \sum_{A=1}^{300} \gamma^A{}_{\mu\nu} R^A ,$$

where R^A and R^I are the projections with anomalous algebra, and the traceless β and γ matrices are defined with respect to the α set through the orthogonality relations

$$\text{Tr}(\alpha^I \beta^J) = \delta^{IJ} \; ; \; \text{Tr}(\alpha^I \gamma^A) = 0 \, .$$

Furthermore, since

$$\begin{aligned}
[M_{\mu\nu}, Q^I] &= 2i\alpha^I{}_{\mu\rho} Q^\rho{}_\nu - (\mu \leftrightarrow \nu) \, , \\
&= 2i \sum_{J=1}^{25} (\alpha^I \beta^J)_{\mu\nu} Q^J + 2i \sum_{J=1}^{25} (\alpha^I \alpha^J)_{\mu\nu} R^J + 2i \sum_{A=1}^{300} (\alpha^I \gamma^A)_{\mu\nu} R^A - (\mu \leftrightarrow \nu) \, ,
\end{aligned}$$

we observe that in general, the action of a Lorentz transformation will produce the forbidden set $\{R\}$. Hence, it is only in the context of a broken Lorentz algebra that the operators Q^I need to be considered. Manifest Lorentz invariance yields an algebra of one extra operator, the BRST of the usual open string field theory. This suggests that we try to arrange the inequivalent sets of α matrices in terms of the unbroken sub-Lorentz algebras. These may be of physical relevance since the world of experience indicates only Lorentz invariance in four dimensions, not in 26 dimensions. Thus only a subgalgebra of the Lorentz algebra will transform the set $\{Q^I\}$ into itself, which then forms a representation (in general reducible) of the unbroken subalgebra. Put differently, under the relevant decomposition of the $SO(25,1)$ Lorentz algebra, $Q_{\mu\nu}$ itself decomposes into a sum of representations of the subalgebra, in terms of which the Q^I must be expressed. This is a powerful requirement which can be used to rule out some imbeddings. For instance, consider the imbedding of F_4 into the Lorentz algebra. It does not work because the symmetric traceless tensor splits up into two F_4 representations, the smallest having 26 dimensions while there are only 25 Q^I's.

In general, the set of α matrices is defined up to linear combinations and an overall similarity transformation, since their defining equation is unaltered by both. We can label the inequivalent sets by the number of elements with non zero determinants, and for the singular matrices by their rank, but this does not shed much light into which Lorentz subgalgebra is left unbroken. The α matrices are of the form

$$\alpha^\mu_\nu = \begin{pmatrix} -\mathrm{Tr}\mathbf{A} & \mathbf{a}^T \\ -\mathbf{a} & \mathbf{A} \end{pmatrix}$$

where $\mathbf{a}$ is a 25-vector and $\mathbf{A}$ is a symmetric 25×25 matrix, obeying the constraints

$$\mathrm{Tr}(\mathbf{A}^I \mathbf{A}^J) + \mathrm{Tr}\mathbf{A}^I \mathrm{Tr}\mathbf{A}^J = 2\mathbf{a}^I \cdot \mathbf{a}^J .$$

We note that if a particular α^I is nilpotent, so is the corresponding operator Q^I, and we can show that there is at most one nilpotent matrix for each set of the matrices. Nilpotency of a given α matrix implies the equations

$$\mathbf{A}^2 = \mathbf{a}\,\mathbf{a}^T$$

$$\mathbf{A}\mathbf{a} = (\mathrm{Tr}\mathbf{A})\mathbf{a} .$$

These equations can be solved by introducing the null 26-dimensional vector a_μ with components $(\mathrm{Tr}\mathbf{A}, \mathbf{a})$, in terms of which we have

$$\alpha_{\mu\nu} = a_\mu a_\nu ; \quad a_\mu a^\mu = 0.$$

Let $\alpha'_{\mu\nu} = b_\mu b_\nu$ be another nilpotent matrix ($b_\mu b^\mu = 0$). We see that $\mathrm{Tr}(\alpha\alpha') = 0$ implies that $a_\mu b^\mu = 0$, so that a_μ and b_μ must be proportional to one another: there is only one nilpotent per set, besides the Lorentz invariant BRST nilpotent.(One can also show that there is no nilpotent linear combination of Q and Q^I; otherwise it would imply

$$\alpha^I(\alpha^I + 1) = 0,$$

meaning that α^I has only 0 or -1 as eigenvalues. Since it is to be traceless, it can only have zero eigenvalues, and thus is similar to a triangular matrix, which contradicts its pseudosymmetry.)

Let us give a specific example. It is very simple and serves to illustrate our procedure which, up to now, has been of a general nature. Let a_μ and b_μ be two null vectors with the properties

$$a^\mu a_\mu = 0 ; \quad b^\mu b_\mu = 0 ; \quad a^\mu b_\mu = 1 ,$$

and introduce a set of 24 transverse orthonormal vectors $c_\mu^{(i)}$ which satisfy

$$c_\mu^{(i)} a^\mu = c_\mu^{(i)} b^\mu = 0 \; ; \; c_\mu^{(i)} c^{(j)\mu} = -\delta^{ij} \; .$$

Then it is easy to see that the set of α matrices can be formed out of these according to

$$\alpha_{\mu\nu}^{(o)} = a_\mu a_\nu \; ,$$

$$\alpha_{\mu\nu}^{(i)} = a_\mu c_\nu^{(i)} + a_\nu c_\mu^{(i)} \; , \quad i = 1, \ldots, 24 \; .$$

This is a rather degenerate example since all these matrices are singular. The ensuing projections, Q^o and Q^i, satisfy the superalgebra

$$\{Q^o, Q^o\} = 0 \; ,$$

$$\{Q^o, Q^i\} = 0 \; ,$$

$$\{Q^i, Q^j\} = -2i\delta^{ij} B^o \; .$$

The algebra can be enlarged to include the BRST charge, Q, with the results

$$\{Q, Q^o\} = 2iB^o \; ,$$

$$\{Q, Q^i\} = 2iB^i \; .$$

The algebra with the Lorentz generators is obtained either directly, or by using the previously outlined method. We see that the orthogonal β and γ matrices are given by

$$\beta_{\mu\nu}^{(o)} = b_\mu b_\nu \; ,$$

$$\beta_{\mu\nu}^{(i)} = b_\mu c_\nu^{(i)} + b_\nu c_\mu^{(i)} \; ,$$

$$\gamma_{\mu\nu}^{+-} = a_\mu b_\nu + a_\nu b_\mu - \frac{1}{13} g_{\mu\nu} \; ,$$

$$\gamma_{\mu\nu}^{ij} = c_\mu^{(i)} c_\nu^{(j)} + c_\mu^{(j)} c_\nu^{(i)} + \frac{1}{13} g_{\mu\nu} \delta^{ij} \; ,$$

respectively (from completeness, $2\gamma^{+-} = \gamma^{ii}$). The relevant projections of the Lorentz generators which map the set into itself are obtained through the antisymmetric light-cone matrices

$$m_{\mu\nu}^{+-} = a_\mu b_\nu - a_\nu b_\mu \; ,$$

$$m_{\mu\nu}^{+i} = a_\mu c_\nu^{(i)} - a_\nu c_\mu^{(i)} \; ,$$

$$m_{\mu\nu}^{-i} = b_\mu c_\nu^{(i)} - b_\nu c_\mu^{(i)} \; ,$$

$$m_{\mu\nu}^{ij} = c_\mu^{(i)} c_\nu^{(j)} - c_\mu^{(j)} c_\nu^{(i)} \; .$$

It is easy to see that the only subalgebra which keeps the set within itself is that generated by the projections $M^{ij} = m^{ij}_{\mu\nu} M^{\mu\nu}$ and $M^{+-} = m^{+-}_{\mu\nu} M^{\mu\nu}$, which together generate the subgroup $SO(24) \times SO(1,1)$, with the transformation properties

$$[M^{+-}, Q^i] = Q^i \; ; \; [M^{+-}, Q^o] = Q^o ,$$

and

$$[M^{ij}, Q^o] = 0 \; ; \; [M^{ij}, Q^k] = i\delta^{ik} Q^j - i\delta^{jk} Q^i .$$

In this very simple example, we see that the algebra preserves Lorentz invariance in $1+1$ dimensions, together with an internal $SO(24)$ algebra, and two nilpotent operators. Note that the set $\{Q^I\}$ forms a reducible representation of the unbroken algebra made up of a singlet Q^o and the Q^i which form a vector representation under the transverse subgroup. The investigation of the representations of this algebra and of the cohomology of the other nilpotent awaits further work[7]. Only then will the physical meaning of this algebra become clear.

It is not very hard, albeit tedious, to produce sets of singular α matrices by inductive procedures. For instance, if we know a set of such matrices in $D-1$ dimensions, we can insert them in a $D \times D$ matrix with the Dth column and row being zeros, and the remaining α matrix will have zeros in its first D rows and columns, and a null vector in its Dth row and column. The eigenvalues of the α matrices are in general complex numbers forming conjugate pairs. In D dimensions, they are determined by $2D-5$ parameters and one overall normalization. There can be at most 7 commuting elements in the non-singular set in 26 dimensions, although this number will increase whenever singular elements are present.

To conclude, we have drawn attention to a curious algebra of normal-ordered, reparametrization invariant operators. We think that these different algebras can be classified in terms of the subalgebra of the Lorentz algebra they contain. In the case of Lorentz invariance in

26 dimensions, this algebra contains only the BRST operator. For Lorentz invariance in 2 dimensions, it contains another nilpotent operator as well as many others. We have not suceeded in presenting intermediate examples, nor have we pursued the physical meaning of these algebras. However, given our state of ignorance in string field theory, it is advisable to study all algebraic structures appearing in the theory.

Acknowledgments

I wish to thank my collaborators G. Kleppe and R. Raju Viswanathan for their participation in this work.

REFERENCES

1. G. Kleppe, P. Ramond, R.R. Viswanathan, *Phys. Lett.* **206B**, 46 (1988); *ibid* Preprint UFTP-88-9, to be published in *Nucl. Phys. B*.
2. P. Ramond, in Proceedings of the First Johns Hopkins Workshop, Jan 1974, G. Domokos and S. Kövesi-Domokos, eds; C. Marshall, P. Ramond, *Nucl. Phys.* **B85**, 375 (1975).
3. E. Witten, *Nucl. Phys.* **276B**, 291 (1986).
4. W. Siegel, *Phys. Lett.* **151B**, 396 (1985).
5. A. Ballestrero and E. Maina, *Phys. Lett.* **180B**, 53 (1986); C. Battle and J. Gomis, Barcelona preprint, 1986 (unpublished); S.P. De Alwis and N. Ohta, *Phys. Lett.* **174B**, 388 (1986); P. Ramond, V.G.J. Rodgers, R.R. Viswanathan, *Nucl. Phys.* **B293**, 293 (1987).
6. E. Witten, *Phys. Rev. Lett.* **61**, 670 (1988).
7. G. Kleppe, P. Ramond, and R.R. Viswanathan, in preparation.

SUPERCONDUCTIVITY OF AN IDEAL CHARGED BOSON SYSTEM

T. D. Lee

Columbia University, New York, N.Y. 10027

ABSTRACT

Recently Friedberg, Lee and Ren have pointed out that at low density the ideal charged boson system turns out *not* to be a superconductor, but becomes a type II superconductor at high density. This conclusion differs from the well-known Schafroth solution of superconductivity at any density for the same problem. Schafroth's analysis is found to contain a mistake due to the neglect of the electrostatic exchange energy E_{ex}. Based on the Schafroth solution, E_{ex} is shown to be $+\infty$ in the normal phase, but 0 in the condensed phase (at $T = 0$). Of course, the correct solution has to give a finite E_{ex}.

This research was supported in part by the U.S. Department of Energy.

1. SCHAFROTH'S SOLUTION

Schafroth's superconductivity solution[1] of an ideal charged boson system published 35 years ago has always been considered to be the definitive work, comparable in depth to the analysis made by Landau[2] on the diamagnetism of an ideal charged fermion system. However, recently it was found[3] that the Schafroth solution contains a serious mistake due to the neglect of the electrostatic exchange energy E_{ex}. It turns out that based on the Schafroth solution, E_{ex} is $+\infty$ in the normal phase, but 0 in the condensed phase (at $T = 0$). Of course, the correct solution has to give a finite E_{ex}.

For ideal charged particles, bosons or fermions, there is only the electromagnetic interaction. These particles are assumed to be nonrelativistic, enclosed in a large volume Ω and with an external background charge density $-e\rho_{ext}$ so that the whole system is electrically neutral; i.e., the integral of the total charge density

$$J_0 \equiv e(\phi^\dagger \phi - \rho_{ext}) \tag{1.1}$$

is zero, where ϕ is the matter field under consideration. In both Landau's and Schafroth's solutions, the expectation value (or the ensemble average, if the temperature $T \neq 0$) of J_0 is zero:

$$< | J_0(\vec{r}) | > = 0, \tag{1.2}$$

except for points near the surface. As $\Omega \to \infty$, this gives a zero Coulomb energy density for a classical system, but not for a quantum system. For the fermions, the electrostatic exchange energy E_{ex} can be calculated, as in the Thomas-Fermi-Dirac model (in contrast to the Thomas-Fermi model); it is not sensitive to the external magnetic field $\vec{H}$. Consequently, for diamagnetism one can ignore E_{ex}. The thermodynamic functions of the N-body system can then be evaluated by using the spectrum of the single-particle Schrödinger equation (in units $\hbar = c = 1$):

$$-\frac{1}{2m}(\vec{\nabla} - ie\vec{A})^2 f(\vec{r}) = = \epsilon f(\vec{r}) \tag{1.3}$$

where $f(\vec{r})$ and ϵ are the eigenfunction and eigenvalue of a single particle of mass m and charge e under a given transverse electromagnetic potential $\vec{A}(\vec{r})$. For fermions, this leads to the well-known Landau formula for the de Haas-van Alphen effect.

On the other hand, for ideal bosons there is the added problem of the Bose-Einstein condensation. As we shall see, the complexity of an ideal charged boson system is largely due to the different manifestations of Coulomb energy E_{Coul} in the two phases. However, in the Schafroth solution for bosons, the electrostatic energy is assumed to depend only on $<|J_0(\vec{r})|>$; hence, as $\Omega \to \infty$ the electrostatic energy density is taken to be zero in both phases (which, as will be analyzed in Section 2, turns out to be a mistake). In the following, we summarize briefly the essential features of the Schafroth solution at $T = 0$.

1.1. Normal Phase in the Schafroth Solution

In the gauge

$$A_y = Bx \quad \text{and} \quad A_x = A_z = 0, \tag{1.4}$$

with B a constant in the interior of Ω, the solution of (1.3) is

$$f_{n,p_y,p_z}(x,y,z) \propto e^{i(p_y y + p_z z)} \psi(x)$$

where $\psi(x)$ satisfies

$$\frac{1}{2m}\left[-\frac{d^2}{dx^2} + (p_y - eBx)^2\right]\psi(x) = (n + \frac{1}{2})\omega\,\psi(x) \tag{1.5}$$

and

$$\omega = eB/m$$

is the cyclotron frequency. To simplify some of the expressions, we adopt the convention

$$e > 0 \quad \text{and} \quad B > 0.$$

Expand the bosonic quantum field operator ϕ in terms of $f_{n,p_y,p_z}(x,y,z)$. At $T = 0$, we can *neglect* the modes $n \neq 0$ and $p_z \neq 0$, and write

$$\phi(x,y,z) = \sum_p a_p \frac{1}{\sqrt{L_y L_z}} e^{ipy} \psi_p(x) \qquad (1.6)$$

where

$$\psi_p(x) = \left(\frac{eB}{\pi}\right)^{\frac{1}{4}} \exp\left[-\frac{1}{2} eB \left(x - \frac{p}{eB}\right)^2\right] \qquad (1.7)$$

which is the solution $\psi(x)$ of (1.5) with

$$n = 0 \quad \text{and} \quad p = p_y. \qquad (1.8)$$

The a_p and its hermitian conjugate $a_p^\dagger$ are annihilation and creation operators, satisfying the commutation relation

$$[a_p, a_{p'}^\dagger] = \delta_{pp'}. \qquad (1.9)$$

The boson density operator is

$$\phi^\dagger(\vec{r})\phi(\vec{r}) = \sum_{p,q} a_p^\dagger a_q \frac{1}{L_y L_z} e^{i(q-p)y} \psi_p(x)\psi_q(x). \qquad (1.10)$$

In the Schafroth solution of the normal phase, because of the zero charge density condition (1.2), the ground state $|\rangle$ is an eigenstate of the occupation number $a_p^\dagger a_p$; i.e.,

$$a_p^\dagger a_p |\rangle = n|\rangle \qquad (1.11)$$

with the same eigenvalue n. (Note that the single particle eigenvalue in (1.3) is $\epsilon = \frac{1}{2}\omega$, independent of p.) Let the volume $\Omega = L_x L_y L_z$ be a rectangular box, and its x-side be within $-\frac{1}{2}L_x$ to $\frac{1}{2}L_x$. The sum over p extends from

$$-\frac{1}{2}eBL_x + O(a) \quad \text{to} \quad \frac{1}{2}eBL_x + O(a) \tag{1.12}$$

in steps of $2\pi/L_y$, with

$$a = (eB)^{-\frac{1}{2}} \tag{1.13}$$

which is the cyclotron radius; hence, the expectation value of (1.10) becomes (for large L_y)

$$<|\phi^\dagger(\vec{r})\phi(\vec{r})|> = \frac{n}{2\pi L_z}\int dp\,[\psi_p(x)]^2. \tag{1.14}$$

It is clear that

$$\int_{-\infty}^{\infty}[\psi_p(x)]^2 dp = eB; \tag{1.15}$$

thus, for x not near the boundary $\pm\frac{1}{2}L_x$, as $L_x \to \infty$

$$<|\phi^\dagger(\vec{r})\phi(\vec{r})|> = N/\Omega \equiv \rho \tag{1.16}$$

provided

$$n = 2\pi N/eB\,L_y L_x = 2\pi L_z \rho/eB \tag{1.17}$$

where N is the total number of particles. Because the constant external density $-\rho_{ext} = -N/\Omega = -\rho$, the expectation value of the charge density $J_0(\vec{r})$ is 0, in accordance with (1.2). Hence, the Schafroth solution gives (in the infinite volume limit) the following energy density for the ground state in the normal phase:

$$\Omega^{-1}E_n = \frac{1}{2}B^2 + \frac{e\rho}{2m}B. \tag{1.18}$$

Now, $\vec{B} \equiv \vec{\nabla} \times \vec{A}$ is the magnetic induction; the magnetic field $\vec{H}$ is parallel to $\vec{B}$, and its magnitude is given by the derivative of $\Omega^{-1}E$ with respect to B:

$$H = B + \frac{e\rho}{2m} \geq \frac{e\rho}{2m} > 0. \tag{1.19}$$

The difference

$$M \equiv B - H = -\frac{e\rho}{2m}$$

is the magnetization.

We assume $L_z \gg L_x$ and L_y (all $\to \infty$ in the end). The boundary condition is

$$\vec{\nabla} \times \vec{A} = H\hat{z} \quad \text{outside } \Omega$$

with $\hat{z}$ the unit vector $\parallel$ the z-axis. In accordance with the Maxwell equation for material, because there is no external current, the magnetic field $\vec{H}$ is uniform throughout the entire space (inside and outside Ω). The magnetic induction $\vec{B} = \vec{\nabla} \times \vec{A} = B\hat{z}$ is uniform only in the interior of Ω; near the surface, $\vec{B}$ varies rapidly from $\vec{B} \neq \vec{H}$ to $H\hat{z}$ on the outside.

Correspondingly, the magnetization $\vec{M} = \vec{B} - \vec{H}$ has a uniform magnitude $-e\rho\hat{z}/2m$ in the interior of Ω, and then changes rapidly across the surface to zero outside Ω; this produces a surface current. At a fixed external field H, the useful free energy is the Legendre transform of the usual Helmholtz free energy F (which at $T = 0$ is the total energy E):

$$\Omega^{-1}\tilde{F} \equiv \Omega^{-1}F - BH. \tag{1.20}$$

For the normal phase, we have, in the infinite volume limit and in terms of H and ρ,

$$\Omega^{-1}\tilde{F}_n = -\frac{1}{2}\left(H - \frac{e\rho}{2m}\right)^2. \tag{1.21}$$

1.2. Critical Field in the Schafroth Solution

In the super phase, it is more convenient to expand the operator $\phi(\vec{r})$ in terms of its Fourier components:

$$\phi(\vec{r}) = \sum_p \Omega^{-\frac{1}{2}} b_{\vec{p}} e^{i\vec{p}\cdot\vec{r}}. \tag{1.22}$$

At $T=0$ all excitations ($\vec{p} \neq 0$) can be neglected. So far as any intensive thermodynamic functions are concerned, we may set, through the spontaneous symmetry-breaking mechanism,

$$\phi = \text{constant} = (N/\Omega)^{\frac{1}{2}} \tag{1.23}$$

and the magnetic induction

$$B = 0 \tag{1.24}$$

in the interior of Ω. Thus, the corresponding E, F and $\tilde{F}$ are given by

$$E_s = F_s = \tilde{F}_s = 0. \tag{1.25}$$

At a given H, $\tilde{F}$ should be a minimum. When H is less than

$$H_c \equiv e\rho/2m, \tag{1.26}$$

there is only the super phase, on account of (1.19). For $H > H_c$, $\tilde{F}_n < \tilde{F}_s$ and therefore the system is in the normal phase. This gives the Schafroth result: the system is a superconductor at any density, provided $H < e\rho/2m$, the critical field.

2. ELECTROSTATIC ENERGY

The total Coulomb energy operator is

$$H_{\text{Coul}} = \frac{e^2}{8\pi} \int\int |\vec{r}-\vec{r}'|^{-1} : J_0(\vec{r})J_0(\vec{r}') : d^3r\, d^3r' \qquad (2.1)$$

where : : (in Wick's notation[4]) denotes the normal product so as to exclude the Coulomb self-energy. In the following, the Schafroth solution will be used to compute[3] the Coulomb energy

$$E_{\text{Coul}} = <|H_{\text{Coul}}|> . \qquad (2.2)$$

We shall show that in the infinite volume limit at $T=0$, $\Omega^{-1}E_{\text{Coul}}$ thus calculated is ∞ in the normal phase, but zero in the super phase.

We first give a general definition of normal versus super phases. Expand the field operator $\phi(\vec{r})$ in terms of any complete orthonormal set of c number (single particle) functions $\{f_i(\vec{r})\}$:

$$\phi(\vec{r}) = \sum_i a_i f_i(\vec{r}) \qquad (2.3)$$

where a_i and $a_i^\dagger$ are operators obeying the commutation relation

$$[a_i, a_j^\dagger] = \delta_{ij} . \qquad (2.4)$$

At a given T, the system is in the normal phase if the canonical ensemble average $<a_i^\dagger a_i>$ for the ith occupation number $a_i^\dagger a_i$ satisfies

$$\lim_{\Omega\to\infty} \Omega^{-1} <a_i^\dagger a_i> = 0 \qquad (2.5)$$

for *any* i and for *any* set $\{f_i(\vec{r})\}$; otherwise, the system is in the super phase. At $T=0$, the ensemble average is the same as the expectation value in the ground state.

In terms of the expansion (2.3),

$$\phi^\dagger(\vec{r})\phi(\vec{r}) = \sum_{i,i'} a_{i'}^\dagger a_i f_{i'}^*(\vec{r})f_i(\vec{r}). \tag{2.6}$$

Consider a normalized state vector $|\ >$ which is an eigenstate of all $a_i^\dagger a_i$:

$$a_i^\dagger a_i |\ > = n_i |\ >; \tag{2.7}$$

hence, the expectation value

$$<|\phi^\dagger(\vec{r})\phi(\vec{r})|> = \sum_i n_i |f_i(\vec{r})|^2, \tag{2.8}$$

$$<|(a_i^\dagger)^2 a_i^2|> = n_i^2 - n_i \tag{2.9}$$

and for $i \neq j$

$$<|a_{i'}^\dagger a_{j'}^\dagger a_i a_j|> = n_i n_j (\delta_{i'i}\delta_{j'j} + \delta_{i'j}\delta_{j'i}). \tag{2.10}$$

Because

$$J_0(\vec{r}) = \phi^\dagger(\vec{r})\phi(\vec{r}) - \rho_{ext},$$

we have

$$\sigma(\vec{r}) \equiv J_0(\vec{r}) - <|J_0(\vec{r})|> \\ = \phi^\dagger(\vec{r})\phi(\vec{r}) - <|\phi^\dagger(\vec{r})\phi(\vec{r})|>, \tag{2.11}$$

$$<|:J_0(\vec{r})J_0(\vec{r}'):|> = <|:\sigma(\vec{r})\sigma(\vec{r}'):|> + <|J_0(\vec{r})|><|J_0(\vec{r}')|> \tag{2.12}$$

and

$$<|:\sigma(\vec{r})\sigma(\vec{r}'):|> = \sum_{i \neq j} n_i n_j f_i^*(\vec{r})f_j(\vec{r})f_j^*(\vec{r}')f_i(\vec{r}') \\ - \sum_i n_i |f_i(\vec{r})|^2 |f_i(\vec{r}')|^2. \tag{2.13}$$

Thus, the expectation value of H_{Coul} can be decomposed into a sum of three terms

$$<|H_{\text{Coul}}|> = E_{\text{ex}} + E_{\text{dir}} + E'_{\text{dir}} \tag{2.14}$$

where

$$E_{\text{ex}} = \sum_{i \neq j} \frac{e^2}{8\pi} \int d^3r\, d^3r'\, |\vec{r} - \vec{r}'|^{-1}\, n_i n_j\, f_i^*(\vec{r}) f_j(\vec{r}) f_j^*(\vec{r}') f_i(\vec{r}'), \tag{2.15}$$

$$E_{\text{dir}} = \frac{e^2}{8\pi} \int d^3r\, d^3r'\, |\vec{r} - \vec{r}'|^{-1}\, <|J_0(\vec{r})|><|J_0(\vec{r}')|> \tag{2.16}$$

and

$$E'_{\text{dir}} = -\frac{e^2}{8\pi} \sum_i \int d^3r\, d^3r'\, |\vec{r} - \vec{r}'|^{-1}\, n_i\, |f_i(\vec{r})|^2\, |f_i(\vec{r}')|^2. \tag{2.17}$$

The direct energy E_{dir} is due to the last term in (2.12); the exchange energy E_{ex} and the other direct (self-) energy E'_{dir} are connected with the first and second terms on the right-hand side of (2.13).

In the super phase, in terms of the Fourier expansion (1.22), all particles are in the $\vec{k} = 0$ state at $T = 0$; therefore, we have, denoting $<|b_{\vec{k}}^\dagger b_{\vec{k}}|>$ by $n_{\vec{k}}$:

$$n_{\vec{k}} = \begin{cases} N & \text{if } \vec{k} = 0 \\ 0 & \text{if } \vec{k} \neq 0. \end{cases} \tag{2.18}$$

This leads to, in the infinite volume limit and at $T = 0$

$$\Omega^{-1} E_{\text{Coul}} = 0 \quad \text{in the super phase}, \tag{2.19}$$

since $<|J_0(\vec{r})|> = 0$, $E_{\text{dir}} = 0$; because of (2.18),

$$E_{\text{ex}} = 0$$

and in addition

$$\Omega^{-1} E'_{\text{dir}} = O(\Omega^{-\frac{1}{3}}) \to 0. \tag{2.20}$$

Lee: Superconductivity of an Ideal Charged Boson System

In the Schafroth calculation, $\Omega^{-1}E_{\text{Coul}}$ is taken to be zero in the normal phase as well. But as we shall see, this is not true if one includes the exchange energy E_{ex} between the bosons.

As mentioned before, in order to have $<|J_0(\vec{r})|>=0$, the Schafroth ground state $|>$ for the normal phase satisfies

$$a_p^\dagger a_p |>\, = n |>\, = \frac{2\pi L_z \rho}{eB} |>, \qquad (2.21)$$

in accordance with (1.17). Thus, in addition to $E_{\text{dir}}=0$, we have, in the infinite volume limit, $\Omega^{-1}E'_{\text{dir}} = O(\Omega^{-\frac{1}{3}}) \to 0$. Hence, the Coulomb energy density $\Omega^{-1}E_{\text{Coul}}$ is determined entirely by the ensemble average of the exchange energy density $\Omega^{-1}E_{\text{ex}}$.

It is useful to introduce

$$D_k(x) \equiv \int dy\,dz\, e^{-ik_1 y - ik_2 z - \mu r}\frac{1}{4\pi r} \qquad (2.22)$$

with k_1, k_2 and μ real, and $\mu > 0$. Since $(-\nabla^2 + \mu^2)(4\pi r)^{-1} = \delta^3(\vec{r})$, we have

$$\left(-\frac{\partial^2}{\partial x^2} + k^2 + \mu^2\right) D_k(x) = \delta(x) \qquad (2.23)$$

where

$$k = (k_1^2 + k_2^2)^{\frac{1}{2}}. \qquad (2.24)$$

Because $D_k(x) \to 0$, as $x \to \infty$, we have

$$D_k(x) = \frac{\ell}{2} e^{-|x|/\ell} \qquad (2.25)$$

with

$$\ell = (k^2 + \mu^2)^{-\frac{1}{2}} > 0.$$

Define

$$d_k(x) \equiv \lim_{\mu \to 0} D_k(x) ; \qquad (2.26)$$

hence,

$$d_k(x) = \frac{1}{2k} \exp(-k|x|) ; \qquad (2.27)$$

with $k > 0$.

Using (1.6), (1.11), (2.15) and (2.27), we find that, to $O(\Omega)$, E_{Coul}, the canonical ensemble average of the Coulomb energy operator, is given by

$$E_{\text{Coul}} = \frac{e^2 n^2}{2 L_y L_z} \sum_{p \neq q} I_{pq} \qquad (2.28)$$

where

$$I_{pq} = \int \int d_{|k|}(x - x') \psi_p(x) \psi_q(x) \psi_p(x') \psi_q(x') \, dx \, dx' \qquad (2.29)$$

and

$$k = p - q \neq 0. \qquad (2.30)$$

Diagramatically, this is the exchange Coulomb energy shown in Figure 1.

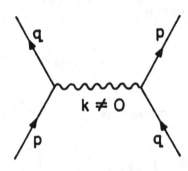

Figure 1

Using (1.7), we can evaluate the integral I_{pq} by first integrating over $\frac{1}{2}(x+x')$, and then $x-x'$:

$$I_{pq} = (\sqrt{2\pi}|k|)^{-1} \int_\tau^\infty e^{-t^2/2}\, dt \qquad (2.31)$$

with $\tau = (eB)^{-\frac{1}{2}}|k|$. Because of (2.21) and as $\Omega \to \infty$, $\sum_p = (2\pi)^{-1} L_y \int dp$ with the range of p varying from $-\frac{1}{2}eBL_x$ to $\frac{1}{2}eBL_x$, we find[3]

$$\Omega^{-1} E_{\text{Coul}} = \frac{1}{2}(e\rho^2/B) \int_{-\infty}^\infty dk\, I_{pq} \qquad (2.32)$$

which diverges logarithmically and *invalidates* the Schafroth solution.

3. CORRECTED SOLUTION

The Coulomb interaction makes the solution of an ideal charged boson system a highly nontrivial many body problem. Because of the complexity of the problem, we restrict our discussions mainly to zero temperature; even so, our analysis is by no means complete.

In the Schafroth solution, the infinite volume limit of $\Omega^{-1}E_{ex}$ in the normal phase is $+\infty$ (at $T=0$). Of course, this situation can be easily remedied by reducing the overlapping of the single particle wave functions between bosons. But, with this done, the boson-density distribution $<|\phi^{\dagger}\phi|>$ would develop fluctuations; hence, E_{dir} and E'_{dir} may both be $\neq 0$.

As is shown in Ref. 3 (and as will also be reviewed in the following), this leads to a solid-like structure in the normal phase at $T=0$. For low density $\rho = N/\Omega$ less than a critical density ρ_c,

$$\rho < \rho_c \sim r_B^{-3}, \tag{3.1}$$

where r_B is the "Bohr" radius defined by

$$r_B^{-1} = e^2 m/4\pi, \tag{3.2}$$

the normal phase has a negative $\Omega^{-1}E_{Coul}$ and is strong enough to make $\tilde{F}_n < 0$; therefore, the ideal charged boson system does *not* exhibit superconductivity. For high density

$$\rho > \rho_c \tag{3.3}$$

(but still $<< m^3$, so that it remains nonrelativistic), the ideal charged boson system is a *type II superconductor*, with H_c much greater than the Schafroth result $e\rho/2m$, given by (1.26).

3.1. Low Density Case

Divide the entire box Ω into N small cubes, each of volume

$$v = \frac{\Omega}{N} \equiv \frac{4\pi}{3} R^3 . \tag{3.4}$$

Take a trial state vector $|\ >$ in which each cube contains one boson, whose single particle wave function $\psi(\vec{r})$ is entirely within that cube. Consequently, the exchange energy (2.15)

$$E_{ex} = 0 . \tag{3.5}$$

It is not difficult to find an upper bound of $E_{dir} + E'_{dir}$, which will turn out to be negative. Approximate the cube (3.4) by a sphere of radius R. The net charge inside each sphere is zero; hence, in the integrals (2.16) and (2.17) for E_{dir} and E'_{dir} we can restrict the integration of $\vec{r}$ and $\vec{r}'$ to the same sphere v:

$$E_{dir} = \frac{e^2}{8\pi} N \int_v |\vec{r} - \vec{r}'|^{-1} j_0(\vec{r}) j_0(\vec{r}') \, d^3r \, d^3r' \tag{3.6}$$

and

$$E'_{dir} = -\frac{e^2}{8\pi} N \int_v |\vec{r} - \vec{r}'|^{-1} |\psi(\vec{r}) \psi(\vec{r}')|^2 , \tag{3.7}$$

where N is the total number of bosons,

$$j_0(\vec{r}) = |\psi(\vec{r})|^2 - \rho_{ext} . \tag{3.8}$$

Set as a trial function

$$\psi(\vec{r}) = (2\pi R)^{-\frac{1}{2}} \frac{1}{r} \sin \frac{\pi r}{R} \tag{3.9}$$

for $r \leq R$, and zero outside. Using (3.5)-(3.8), we find

$$E_{Coul} = \frac{e^2}{2} N \rho_{ext} \int_0^R 4\pi r^2 V \, dr$$

where $V(r)$ is the electric potential generated by a charge density $-2|\psi(\vec{r})|^2 + \rho_{\text{ext}}$ within the sphere of radius R. This gives an approximate upper bound

$$E_{\text{Coul}} < -\frac{e^2 N}{4\pi R} K \qquad (3.10)$$

where

$$K = \frac{11}{15} + \frac{1}{4\pi^2}. \qquad (3.11)$$

Correspondingly, there is a kinetic energy (in the absence of a magnetic field)

$$E_{\text{kin}} = \frac{\pi^2}{2mR^2} N.$$

The sum $E_{\text{Coul}} + E_{\text{kin}}$ is negative if

$$\rho^{\frac{1}{3}} < 2 \left(\frac{3}{4\pi}\right)^{\frac{1}{3}} \frac{K}{\pi^2} r_B^{-1}. \qquad (3.12)$$

Thus, an approximate lower bound for the critical density ρ_c is

$$\rho_c > \left(\frac{2K}{\pi^2}\right)^3 \frac{3 r_B^{-3}}{4\pi}. \qquad (3.13)$$

Of course, one can readily improve this lower bound by using better trial functions. In the low density case, write

$$E_{\text{kin}} + E_{\text{Coul}} = -N \epsilon_b, \qquad (3.14)$$

where $\epsilon_b > 0$ is the binding energy per particle in the normal phase. The system is a solid at low temperature.

For weak magnetic field, there is in addition a diamagnetic energy E_{dia} which can be calculated perturbatively:

$$E_{\text{dia}} = \frac{e^2}{8m} \overline{r^2} B^2 N, \qquad (3.15)$$

where $\overline{r^2}$ is the mean squared radius of $|\psi(r)|^2$. At $T=0$, the Helmholtz free energy density $\Omega^{-1}F$ is the same as the total energy density, which for the normal phase is

$$\Omega^{-1}F_n = \Omega^{-1}E_n = -\rho\epsilon_b + \frac{1}{2}B^2 + \frac{1}{2}\nu B^2 \qquad (3.16)$$

where $\nu = e^2\overline{r^2}/4m$. Differentiating (3.16), we find $H = B(1+\nu)$ and

$$\tilde{F}_n = -\rho\epsilon_b - \frac{1}{2}(1+\nu)^{-1}H^2. \qquad (3.17)$$

From the curve F versus B, one can construct $\tilde{F} = F - BH$ graphically. At any point P on the curve $F(B)$, draw a tangent. Since the slope $\partial F/\partial B$ is H, the intercept of the tangent to the ordinate gives $\tilde{F} = F - BH$, as shown in Figure 2. Due to stability, $\partial^2 F/\partial B^2 = \partial H/\partial B > 0$. From Figure 2, one sees that if the normal phase has a negative Helmholtz free energy F_n at $B=0$, then $\tilde{F}_n$ is negative at any B (therefore also at any H).

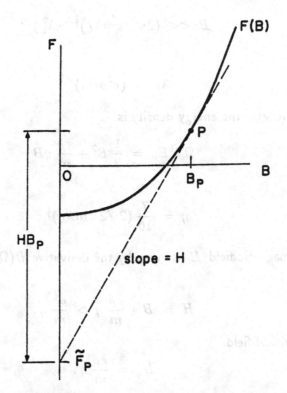

FIGURE 2.

However, the corresponding thermodynamic energy density for the super phase is (as $\Omega \to \infty$)

$$F_s = \tilde{F}_s = 0.$$

Since $\tilde{F}_n < 0$, the system is *not* a superconductor when ρ is $< \rho_c$.

3.2. High Density Case

The high density case is quite complicated. Here, we list only the main results derived in Ref. 3.

A. Normal phase

By using trial functions, the following *upper* bounds are found for the energy density $\Omega^{-1} E_n$ of the normal phase at $T = 0$, $\rho > \rho_c \sim r_B^{-3}$ and in the infinite volume limit:

(i) Weak field

For

$$B \ll (2\sqrt{2}\,\pi m \lambda_L)^{\frac{2}{3}} (e\lambda_L^2)^{-1} \tag{3.18}$$

where

$$\lambda_L = (e^2 \rho/m)^{-\frac{1}{2}} \tag{3.19}$$

is the London length, the energy density is

$$\Omega^{-1} E_n = \frac{1}{2} B^2 + \frac{e\rho}{m} \eta B \tag{3.20}$$

where

$$\eta = \frac{7}{16} (2\sqrt{2}\,\pi m \lambda_L)^{\frac{2}{3}}. \tag{3.21}$$

The external magnetic field H is given by the derivative $\partial(\Omega^{-1} E_n)/\partial B$:

$$H = B + \frac{e\rho}{m} \eta > \frac{e\rho}{m} \eta. \tag{3.22}$$

This gives a critical field

$$H_c = \frac{e\rho}{m} \eta. \tag{3.23}$$

Since $\eta \gg 1$, H_c is much greater than the Schafroth result $e\rho/2m$. (Of course, the Schafroth solution would also give an incorrect $\Omega^{-1} E_n = \infty$.)

(ii) Intermediate field

For
$$(\rho m)^{\frac{1}{2}} \gg B \gg (2\sqrt{2}\,\pi m \lambda_L)^{\frac{1}{3}}(e\lambda_L^2)^{-1}, \tag{3.24}$$

the energy density is
$$\Omega^{-1} E_n = \frac{1}{2} B^2 + c\rho^{\frac{5}{6}} r_B^{\frac{4}{3}}(eB)^{\frac{3}{2}} \tag{3.25}$$

where
$$c = 2^{-5}(3\pi)^{-\frac{2}{3}} 5 e^2. \tag{3.26}$$

(iii) Strong field

For
$$B \gg (\rho m)^{\frac{1}{2}}, \tag{3.27}$$

we find
$$\Omega^{-1} E_n = \frac{1}{2} B^2 + \frac{e\rho}{2m} B + \frac{e\rho^2}{B} \gamma_n \tag{3.28}$$

where
$$\gamma_n = 0.01282. \tag{3.29}$$

B. Super Phase

For the super phase we assume that at $T = 0$ the ground state wave function is
$$<\vec{r}_1, \cdots, \vec{r}_N| > = \prod_{i=1}^{N} \phi_c(\vec{r}_i); \tag{3.30}$$

i.e., the quantized field operator $\phi(\vec{r})$ can be approximated by a classical field $\phi_c(\vec{r})$. We list only the energy density $\Omega^{-1} E_s$ of the super phase at $T = 0$ and in the infinite volume limit.

(i) Weak field

For a weak external magnetic field H,

$$H < H_{c_1} \tag{3.31}$$

where

$$H_{c_1} = \frac{e\rho}{2m}\left[\ln(\lambda_L/\xi) + 1.623 + O(\xi/\lambda_L)\right], \tag{3.32}$$

and ξ is the coherence length given by

$$\xi = (2\lambda_L/m)^{\frac{1}{2}} \ll \lambda_L, \tag{3.33}$$

we have

$$B = 0 \quad \text{and} \quad \Omega^{-1}E_s = 0. \tag{3.34}$$

(ii) Intermediate field

For $H > H_{c_1}$, but

$$H \ll (\rho m)^{\frac{1}{2}}, \tag{3.35}$$

vortex filaments appear in the super phase, with

$$\Omega^{-1}E_s = \frac{1}{2}B^2 + \frac{e\rho}{2m}B\left[-\frac{1}{2}\ln(2\pi e B\xi^2) + \text{constant}\right]. \tag{3.36}$$

The constant is 2.034 for a square lattice and 2.024 for a regular triangular lattice (neglecting $O(\xi/\bar{\ell})$ and $O(\xi/\lambda_L)$).

(iii) Strong field

For

$$H \gg (\rho m)^{\frac{1}{2}}, \tag{3.37}$$

we can use the Abrikosov solution[5] to evaluate the Coulomb energy as a perturbation. This gives

$$\Omega^{-1} E_s = \frac{1}{2} B^2 + \frac{e\rho}{2m} B + \frac{e\rho^2}{B} \gamma_s \qquad (3.38)$$

with

$$\gamma_s = 0.01405 \quad \text{for a square lattice} \qquad (3.39)$$

and

$$\gamma_s = 0.01099 \quad \text{for a regular triangular lattice.} \qquad (3.40)$$

From (3.29), we see that

$$\gamma_n = 0.01282 > \gamma_s = 0.01099.$$

Since both are only upper bounds based on trial functions, this does not mean the correct γ_n is greater than the correct γ_s. But in the effort to derive good upper bounds (3.29) and (3.40) for γ_n and γ_s, it seems probable to suppose that these upper bounds are either very near, or equal to, the exact values. If indeed

$$\gamma_n > \gamma_s, \qquad (3.41)$$

then at $T = 0$

$$H_{c_2} = \infty, \qquad (3.42)$$

which means for $\rho > \rho_c \sim r_B^{-3}$, the ideal charged boson system is always in the super phase.

At any T, besides the critical field $H_c(T)$, there are three critical magnetic fields (arranged in ascending order of their strength):

$$H_{c_1} \sim \frac{e\rho_c}{2m} \ln \lambda_L/\xi,$$

$$\qquad (3.43)$$

$$H'_{c_2} \sim \sqrt{\rho_c m} \sim (e\xi^2)^{-1}$$

and H_{c_2}, where λ_L and ξ are the London length and the coherence length at T:

$$\lambda_L^{-2} = e^2 \rho_c/m \quad \text{and} \quad \xi = (2\lambda_L/m)^{\frac{1}{2}}$$

with $\rho_c =$ condensate density. The "cyclotron" radius $a \equiv (eH)^{-\frac{1}{2}}$ is slightly smaller than λ_L at $H \sim H_{c_1}$, and $\sim \xi$ when $H \sim H'_{c_2}$, but $\ll \xi$ for $H \gg H'_{c_2}$ (so is, therefore, the average distance between vortex filaments). For $H > H_{c_2}(T)$, the system is in the normal phase.

For the ideal charged fermions, there is a relatively small Debye length, which screens the Coulomb interaction. For the ideal charged bosons, at low temperature the Coulomb forces make the normal phase behave very differently from a gas; rather, it resembles a solid or a liquid. Consequently, there is no effective means of screening. (For practical applications, one has to extend the model to include electrons or holes. In that case, the Debye length of the fermions can provide Coulomb screening, as discussed in the boson-fermion model[6] recently proposed for high T_c superconductivity.)

REFERENCES

[1]. M.R. Schafroth, Phys.Rev. **100**, 463 (1955).

[2]. L.D. Landau and E.M. Lifshitz, *Statistical Physics*, Part 1 (Pergamon Press, 1980), pp. 171-177.

[3]. R. Friedberg, T.D. Lee and H.C. Ren, "A Correction to Schafroth's Superconductivity Solution of an Ideal Charged Boson System," Preprint CU-TP-460.

[4]. G.C. Wick, Phys.Rev. **80**, 268 (1950).

[5]. A.A. Abrikosov, Soviet Phys. JETP **5**, 1174 (1957).

[6]. R. Friedberg, T.D. Lee and H.C. Ren, "Coherence Length and Vortex Filament in the Boson-Fermion Model of Superconductivity," Preprint CU-TP-457.

Some Remarks on the Symmetry Approach to Nuclear Rotational Motion*

L. C. BIEDENHARN

Department of Physics, Duke University

Durham, NC 27706 USA

and

PIERO TRUINI

Università di Genova, Istituto di Fisica

Genova, Italy

ABSTRACT:

We discuss how Murray Gell-Mann contributed to the theory of nuclear rotational motion.

---

* Supported in part by the DOE and the NSF

1. Introduction

Nuclear physics could hardly be called one of Murray Gell-Mann's primary interests, but it is our aim in the present note to show, nonetheless, that Murray did in fact make basic contributions to the theory of nuclear rotational motion. Of course—from the perspective of the next century—Murray Gell-Mann's introduction of quarks and SU_3(color) will be seen as laying the very foundations of theoretical nuclear physics itself, so it will not be surprising to anyone (in that era) that he contributed to nuclear rotational theory.

Murray Gell-Mann developed his ideas, which we will discuss below, in the 1960's entirely in the context of particle physics and, although aware of their importance for other fields, he did not himself publish applications outside particle physics or field theory. By good fortune one of us (LCB) was a visiting faculty member at Cal Tech at this critical time and one day an invitation came to visit his office. During this visit [a] he explained at some length the significance of his approach for nuclear physics, and for nuclear rotational motion in particular, and suggested that these ideas be followed up. It was in this way that Murray's ideas made their way into the nuclear physics literature.

2. The Gell-Mann Approach

The essence of Gell-Mann's approach to nuclear rotational motion is to focus attention on the operators that are to characterize the physics and to extract physical results from the algebra these operators generate. For rotational motion these operators include the angular momentum, (**J**), and the nuclear quadrupole operator, (**Q**), which measures (roughly) the shape of the nucleus. There are many other possible operators that could be relevant (octupole operators, for example) but these two capture the essentials of the problem.

---

[a] The date can be recalled easily since—in typical Gell-Mann fashion!—he pointed out that it was Boxing Day, (26 December 1968), a day, which he explained, is celebrated in England not for pugilism but for charity by the royal family.

But how can one do physics with just these two operators? Real physics involves dynamics, and it is at this point Gell-Mann introduced (in particle physics) a qualitatively different approach to symmetry.[1] Prior to this work, the customary (Wigner) approach to symmetry consisted in treating the Hamiltonian for a system as having a dominant term, which respects a given symmetry group, and a small term which perturbs this symmetry. The approach thus assumes the existence of a small parameter whose vanishing leads to complete symmetry. The pattern of the multiplet splittings and the ratios of transition strengths are then group theoretically determined, depending only on the particular group at hand.

Gell-Mann introduced a very different procedure. Following his viewpoint, we may say that we do not care if a symmetry breaking is large or small— provided only that the Hamiltonian be a function of certain *transition operators* which generate the symmetry. This requirement alone suffices to ensure that any multiplet characteristic of the symmetry will at worst be split by the Hamiltonian, *but not mixed with other multiplets.*

This approach starts then with a set of operators that obey (the equal time) commutation relations characteristic of some algebra. These operators are identified with physical transition operators, which when acting on a given state, use up most of their strength in transitions to a few nearby states. The algebra may be such that (because of dynamics) the stationary, or quasi-stationary states, fall into a few (unitary) irreducible representations of the group. If so, this is said to be a "good symmetry."

In addition to emphasizing the changed rôle played by symmetry, Gell-Mann also emphasized a second point: that *commutation relations are kinematical statements*, so that the algebraic structure is preserved independently of symmetry breaking. Following the model by which the weak and electromagnetic currents were exploited, Gell-Mann[b] considered the (symmetric) energy-momentum tensor, which couples to gravity, and showed

---

[b] (together with his collaborators Y. Dothan and Y. Ne'eman)

that in a simple Lagrangian quark field theory model the time derivative of the quadrupole moment of the energy and the orbital angular momentum close on the algebra of $SL(3, R)$. This is the group of volume preserving deformations and rotations of three-space.

Let us emphasize the interesting way in which the algebra of $SL(3, R)$ was obtained. The method is actually an example of an earlier technique Gell-Mann had invented, which incorporates an important new idea: "anti-contraction". The idea of contraction is one that physicists have long been familiar with, and can best be explained by example. Consider relativistic symmetry (the Poincaré group) and take the limit in which the velocity of light becomes infinite. Then the Poincaré group goes over into non-relativistic Galilei group. Gell-Mann's ingenious idea is a way to go *backwards*, that is, to undo the limit: this is "anti-contraction".

For the nuclear physics case, the trick was accomplished by using the Hamiltonian. The relevant part of the Hamiltonian is, in fact, just proportional to $\mathbf{J} \cdot \mathbf{J}$. Thus we define a new operator: (call it $\mathbf{T}$) the time derivative of the quadrupole operator $\mathbf{Q}$:

$$\mathbf{T} \equiv i[H, \mathbf{Q}] \cong i[\mathbf{J} \cdot \mathbf{J}, \mathbf{Q}]. \qquad (1)$$

The miracle of anti-contraction now happens: the new operator $\mathbf{T}$ and the angular momentum operator $\mathbf{J}$ now have non-trivial commutation relations (most importantly $[\mathbf{T},\mathbf{T}] \subset \mathbf{J}$) and close on the Lie algebra of $SL(3, I\!R)$. In the limit that the nuclear volume is made increasingly stiff toward deformations, the $SL(3, I\!R)$ group *contracts* to the original algebra where the quadrupole operator has *commuting* components.

The Gell-Mann approach to symmetry was applied to nuclear physics in[2] and the basic model assumption, $T = i[H, Q]$, was tested by comparing experimental $E2$ transitions [taking the electric- and mass-quadrupole operators to be proportional] for rotational nuclei. [The model predicts ratios of intraband and interband $E2$ transitions with *no ad-*

justable parameters.] The results are quite favorable, and give clear evidence that $SL(3,I\!R)$ symmetry is important in nuclear physics.

The second prediction from the model, quadrupole sum rules, has been examined by Bohr,[3] who finds that roughly 50% of the $E2$ strength is accounted for. Thus $SL(3,I\!R)$ though useful, is not (as one might expect) the whole story.

It is fundamental aspect of rotational motion that because of the quantization of angular momentum an arbitrarily small rotational frequency cannot be assumed for a rotating system. It follows that the adiabatic splitting of the wavefunction (as in the original Bohr-Mottelson model) into a product of a fixed intrinsic wavefunction and a rotational wavefunction (giving the orientation) cannot be a general property. An alternative way to view the physics of this situation is to note that the adiabatic splitting implies that the relation between the body-fixed frame (the frame of the intrinsic wavefunction) and the laboratory frame is well defined. But to define the angular variables relating to the two frames implies by the uncertainty principle that large angular momenta are involved. That is to say, the intrinsic wavefunction is required to be essentially unchanged even for large rotational excitations. Using still other words, the adiabatic splitting necessarily implies that the rotational bands (effectively) *do not terminate.*

Unlimitedly large (rotational) bands for example are characteristic of noncompact groups such as $SL(3,I\!R)$ (and infinite-dimensional unitary representations). By contrast, the nuclear $SU3$ model concerns a compact group and the associated band structure *must terminate.*

There is an interesting way to view the structural differences between terminating and non-terminating bands. Let us consider the quadratic invariant (Casimir) operator, I_2 for the relevant groups: ($L=$ angular momentum, $Q=$ quadrupole generator).

Compact group (Elliott $SU3$): $I_2 = \langle L^2 \rangle + \langle Q:Q \rangle$.

Noncompact group (Gell-Mann $SL(3,I\!R)$): $I_2 = -\langle L^2 \rangle + \langle Q:Q \rangle$.

One sees that the invariant, I_2, defines a "conic section" (quadratic form) in the variables L and Q. It follows that for the compact case the *quadrupole moment* must cut-off with increasing L; this is the elliptic case. By contrast, the noncompact case (hyperbolic case), the quadrupole moment not only does not break-off but gets increasingly *large!*

(There is an intermediate case implied by the analogy to the conic sections: this is the parabolic case which corresponds to the invariant: $I_2 = \langle Q : Q \rangle$. This group corresponds to a *constant* quadrupole moment, independent of the angular momentum.)

It is an essential feature of the noncompact group structure that the non-compact operator moment in $SL(3, \mathbb{R})$ increases with angular momentum. If this moment is related to the electric quadrupole transition operator then the $B(E2)$ value (cross-section) necessarily becomes arbitrarily large. This is in clear contradiction to experiment! Before one jumps to the conclusion that a noncompact group (such as $SL(3, \mathbb{R})$) is forthwith excluded, let us note how the Gell-Mann structure for $SL(3, \mathbb{R})$ escapes this dilemma: the non-compact generator is now the *time-derivative* of the physical (electric charge) quadrupole operator. Thus the increase in the size of the (electric) quadrupole operator is compensated by the increase in the transition energy. (In fact, if one takes the energy spectrum of a rigid rotator, the (Gell-Mann) $B(E2)$ moment becomes that of the rigid rotator.)

We may conclude from this discussion that the $SL(3, \mathbb{R})$ approach to nuclear collective motion is *very tightly structured indeed* and the Gell-Mann anti-contraction procedure is a most essential (and physical) structural element.

3. Some Irreps of $SL(3, \mathbb{R})$

Once one has identified the group $SL(3, \mathbb{R})$ as the proper symmetry structure for nuclear rotational motion, then it becomes of interest to determine the (unitary) irreducible representations (irreps) of this group since these irreps are actually a complete listing of all possible models for physical nuclei, to the extent that the $SL(3, \mathbb{R})$ symmetry is itself valid.

Since $SL(3,\mathbb{R})$ is a rank two group, the irreps are labelled by the eigenvalues of two invariant operators. Adopting the notation used for $SU(3)$ we denote the $SL(3,\mathbb{R})$ irreps by $[p,q,0]$ where now p,q may be complex.

The simplest irreps are those having the labels $[-3/2+i\eta,0,0]$, η real. There are two such irreps each corresponding to a single band:

$$0^+: \quad L = 0,2,4,\ldots,$$

$$0^-: \quad L = 1,3,5,\ldots.$$

The quantum mechanics of the rigid axially symmetric top shows that these bands have for the K-quantum number $K = 0$, and are split by reflection symmetry into odd and even parity, as indicated. The Casimir invariant I_2 has the eigenvalue $I_2 = \frac{1}{8} + \frac{1}{18}\eta^2$ (η real) and this continuous parameter is related to the moment of inertia.

There exists a family of irreps having two continuous labels. The most interesting for physical applications are the sub-set characterized by the finite subgroup V (= Vierergruppe) having representation labels (++).

The angular momentum content for this latter case is found to consist of an infinite series of K-bands:

$$K = 0: \quad L = 0,2,4,\ldots,$$
$$K = 2: \quad L = 2,3,4,\ldots,$$
$$K = 4: \quad L = 4,5,6,\ldots,$$
$$\cdots$$

The labels $[p,q,0]$ are now: $p = \frac{\sigma-3}{2} + i\eta$; $q = \sigma > 0$; σ, η real numbers.

The resemblance of this family of irreps to solutions of the Bohr Hamiltonian[3] is noteworthy.

An important extension of these ideas was obtained in Refs. (2,4) in which it was shown that it is the (two-fold) covering group $\overline{SL(3,\mathbb{R})}$ which is physically significant, and that one obtains half-integer, as well as integer, bands in this way.

The two-fold covering group $\overline{SL(3,\mathbb{R})}$ brings with it a surprise: there is an irrep (which Gell-Mann called *the quarkel*) with the spin content:

$$\text{quarkel irrep:} \quad J = 1/2, 5/2, 9/2, \ldots .$$

The quarkel is the most elementary mode possible for nuclear rotational motion, and can in fact be used as the basic building block to construct all $\overline{(SL(3,\mathbb{R})}$ irreps.

(Yuval Ne'eman has found an interesting use for the quarkel as a global spinor for general relativity. Cartan's theorem which forbade the existence of global spinors in relativity explicitly excluded (as Ne'eman observed) *finite* dimensional spinors and not infinite dimensional spinors such as the quarkel.)

There is an interesting relation between the irreps of this group and the shell model (more precisely, the shell model based on the harmonic oscillator without spin-orbit coupling). Usually one considers the states ordered by the number of quanta: $N = 0\hbar\omega$, $L = 0$; $N = 1\hbar\omega$, $L = 1$; $N = 2\hbar\omega$, $L = 0, 2$; etc. The states belonging to $N\hbar\omega$ are precisely the $SU(3)$ irreps [N,0,0] in an (orbital) angular momentum decomposition (degenerate in energy since $SU(3)$ is the symmetry group of this shell model). If we plot these states in an energy (N) vs angular momenta (L) diagram, the $SU(3)$ irreps are *horizontal* bands.

The states associated with irreps of $SL(3,\mathbb{R})$ are these same states but this time organized *diagonally*. States along the lowest diagonal ($N = 0$, $L = 0$; $N = 1$, $L = 1$; $N = 2$, $L = 2$; $\cdots$) form an infinite array which consists of two irreps of the group $SL(3,\mathbb{R})$, the two $K = 0$ irreps $[-3/2 + i\eta, 0, 0]$. The (N, L) plot shows that every diagonal has, in fact, the same set of two irreps.

Thus the states belonging to the (orbital) shell model may be organized into sets in two ways by the same complex Lie algebra A_2:

(i) Horizontally = compact Lie group $SU(3)$, degenerate multiplets typifying a symmetry group, or

(ii) Diagonally = non-compact group $SL(3, \mathbb{R})$, typifying a spectrum-generating group.

4. Other ways to introduce $SL(3, \mathbb{R})$

It is interesting that there are other, very different, ways in which the same group $SL(3\mathbb{R})$ may be introduced into nuclear structure. If we take (as a rough physical approximation) the energy of a nucleus to be largely independent of shape and hence a function only of volume—then we find that the symmetry group for such a structure is that of volume conserving deformations and rotations: this is once again the group $SL(3, R)$.

There is an alternative way[5] to introduce the group $SL(3, \mathbb{R})$ by considering coherent linear motion for all the particles in the nucleus. That is, one uses:

$$x_{in}(t) = M_{ij}(t) x_{jn}(0), \tag{2}$$

where $\{x_{in}\}$ are the coordinates (i=1,2,3) of the nth nucleon. To preserve volume one imposes the restriction: det $M = 1$. In this way one arrives at $SL(3, \mathbb{R})$—the group of 3×3 real matrices with unit determinant having matrix products as the group law. Every such matrix may be factored in the form:

$$M = R_1 \Delta R_2, \tag{3}$$

where R_1 and R_2 are independent 3×3 rotation matrices, and Δ is a diagonal matrix of determinant unity.

This factorization is important since it shows that the elements of $SL(3, \mathbb{R})$ are:

(i) coordinate frame rotations (R_1),

(ii) vortex spin rotations (R_2), and

(iii) constant volume deformations (Δ).

It is in this way that one can identify an important feature of $SL(3, I\!R)$: the existence of a basic physical element in nuclear collective motion, the *vortex spin*.

Note that the independent existence of vortex spin rotations requires the existence of deformations ($\Delta = 1$ implies R_1 and R_2 are not distinct).

Let us recall the two-fold spin covering, which allows the existence of half-integer spins. This is important for understanding vortex spin, as we shall discuss below.

5. A microscopic basis for the vortex spin of $\overline{SL(3, I\!R)}$

The existence of integer and half-integer vortex spin in of $\overline{SL(3, I\!R)}$ symmetry poses an important physical question: can one give a microscopic basis for the vortex spin?

Let us denote the Cartesian coordinates for a system of A interacting (non-relativistic) nucleons in three-dimensional Euclidean space by:

$$\{x_{in}; \; i = 1, 2, 3, \; n = 1, 2, \ldots A\} \in I\!R^{3A}. \tag{4}$$

The $\{x_{in}\}$ comprise $3A$ variables that certainly must enter in the microscopic description of the nuclear motion, but what should one use as the collective coordinates? This question has a long and involved history—much too difficult for us to treat adequately here—but the answer[1)-3)] is remarkably simple.[6,7,8] (For the moment let us omit discussing the center of mass coordinates.)

There are two symmetric tensors that are naturally associated with the $\{x_{in}\}$:

(a) the mass-quadrupole tensor: $\quad Q_{ij} \equiv \sum_{n=1}^{A} x_{in} x_{jn},$ \hfill (5)

(b) the tensor in "particle label space": $\quad \mathcal{Q}_{nn'} \equiv \sum_{i=1}^{3} x_{in} x_{in'}.$ \hfill (6)

It is equally natural—since these two tensors are real and symmetric—to bring them to diagonal form. It is a pleasant surprise that (for $A \geq 3$), the non-zero eigenvalues of $\mathcal{Q}$ are identical to three eigenvalues of $\mathbf{Q}$. For the mass-quadrupole tensor we may write:

$$\mathbf{Q} = \sum_{\alpha=1}^{3} \mathbf{s}_\alpha \lambda_\alpha \mathbf{s}_\alpha, \quad (\mathbf{s}_\alpha \cdot \mathbf{s}_\beta) = \delta_{\alpha\beta}, \tag{7}$$

where the $\{\mathbf{s}_\alpha\}$ are orthonormal eigenvectors which define a *body-fixed* or *intrinsic* reference frame and the $\{\lambda_\alpha = \sum_{n=1}^{A}(x_{\alpha n})^2\}$ are the three positive eigenvalues. For the tensor in label space one finds:

$$\mathcal{Q} = \sum_{\alpha=1}^{3} \mathbf{v}_\alpha \lambda_\alpha \mathbf{v}_\alpha, \quad (\mathbf{v}_\alpha \cdot \mathbf{v}_\beta) = \delta_{\alpha\beta}. \tag{8}$$

The eigenvectors of $\mathcal{Q}$ belonging to zero eigenvalue form an $A-3$ dimensional degenerate subspace embedded in A dimensions and they may be chosen in many different ways to be orthonormal among themselves and orthogonal to the three distinguished eigenvectors $\mathbf{v}_\alpha$ belonging to eigenvalues λ_α. These $A-3$ vectors will be denoted by $\mathbf{v}_\kappa$ ($\kappa \neq 1, 2, 3$) and referred to as comprising the *zero* subspace for which $\mathbf{v}_\kappa \cdot \mathbf{v}_{\kappa'} = \delta_{\kappa\kappa'}$ and $\mathbf{v}_\kappa \cdot \mathbf{v}_\alpha = 0$. The three distinguished vectors $\mathbf{v}_\alpha$ will be said to span the *internal* subspace.

To avoid square roots it is helpful to introduce the new collective variables $\{\mu_\alpha\}$, defined by

$$\mu_\alpha \equiv +(\lambda_\alpha)^{1/2} \quad (\alpha = 1, 2, 3). \tag{9}$$

(This is allowed since the λ_α are never negative.)

It can be shown[8] that one may write the microscopic coordinates in the form:

$$\{x_{in}\} = \sum_{\alpha=1}^{3} \mathbf{v}_\alpha \mu_\alpha \mathbf{s}_\alpha. \tag{10}$$

This result expresses the microscopic coordinates in an elegantly simple intrinsic way: the three internal orthonormal eigenvectors $\{v_\alpha\}$ of the particle label tensor $\mathcal{Q}$ and the three orthonormal principal axis eigenvectors $\{s_\alpha\}$ of the coordinate space tensor $\mathbf{Q}$ are linked with the three collective coordinates $\{\mu_\alpha\}$ which are square roots $\mu_\alpha = (\lambda_\alpha)^{1/2}$ of the common eigenvalues λ_α of $\mathcal{Q}$ and $\mathbf{Q}$. One could hardly have expected *a priori* such a simple and transparent result.

The removal of the center-of-mass collective coordinate can be shown to be the replacement: $A \to A - 1$, with the $\{v_\alpha\}$ orthogonal to the center of mass vector $v_0 = (1, 1, \ldots, 1)$, which belongs to the zero subspace. Making this change the transformation to collective plus internal coordinates corresponds to: 3 center-of-mass coordinates, 3 Euler angles (from the s_α), 3 vibrational coordinates (the μ_α)—a total of 9 *collective coordinates*—with the remaining $3A - 9$ coordinates being the *intrinsic coordinates*—belonging to the coset manifold $\mathcal{O}(A-1)/\mathcal{O}(A-4)$—which determine the $\{v_\alpha v_\kappa\}$.

The quantal kinetic energy operator T_{qu} can be similarly transformed:

$$T_{qu} = \frac{\mathbf{P}^2}{2MA} + \sum_\alpha \frac{p_\alpha^2}{2M}$$
$$+ \sum_{\alpha<\beta} \frac{\lambda_\alpha + \lambda_\beta}{2M(\lambda_\alpha - \lambda_\beta)^2} L_{\alpha\beta}^2 + \sum_{\alpha<\beta} \frac{\lambda_\alpha + \lambda_\beta}{2M(\lambda_\alpha - \lambda_\beta)^2} \mathcal{L}_{\alpha\beta}^2 + \sum_{\alpha\kappa} \frac{\mathcal{L}_{\alpha\kappa}^2}{2M\lambda_\alpha}$$
$$+ \sum_{\alpha<\beta} \frac{4\mu_\alpha\mu_\beta}{2M(\lambda_\alpha - \lambda_\beta)^2} L_{\alpha\beta}\mathcal{L}_{\alpha\beta}, \tag{11}$$

where $\mathbf{P}$ is the center-of-mass momentum operator, $p_\alpha \equiv -i\hbar\partial/\partial\mu_\alpha$ is the vibrational momentum operator, $\{\mathcal{L}_{\alpha\beta}, \mathcal{L}_{\alpha\kappa}\}$ generate rotations in the internal space, and $\{L_{\alpha\beta}\}$ is the angular momentum referred to the body-fixed frame.

The quantal kinetic energy operator thus splits into a sum of three contributions: a *collective part* (first three terms in eq. (11)), an *internal part* (next two terms) and a *coupling term* (last term). The origin of the coupling term is the linkage between the collective and internal spaces through the common body-fixed distinguished reference frame.

This geometric approach identifies *two commuting angular momenta*:

$$L_\gamma \equiv (\mathbf{L}\cdot\mathbf{s}_\gamma) = M\sum_{n=1}^{N}(\mathbf{r}_n\times\dot{\mathbf{r}}_n)\cdot\mathbf{s}_\gamma, \quad (\alpha\beta\gamma \text{ cyclic with values } 1,2,3)$$

$$= M\sum_{n=1}^{N}\{(\mathbf{s}_\alpha\cdot\mathbf{r}_n)(\dot{\mathbf{r}}_n\cdot\mathbf{s}_\beta) - (\mathbf{s}_\beta\cdot\mathbf{r}_n)(\dot{\mathbf{r}}_n\cdot\mathbf{s}_\alpha)\} \equiv L_{\alpha\beta}, \qquad (12)$$

$$\mathcal{L}_{\alpha\beta} \equiv -M\sum_{n=1}^{N}\left\{\left(\frac{\mu_\beta}{\mu_\alpha}\right)(\mathbf{s}_\alpha\cdot\mathbf{r}_n)(\dot{\mathbf{r}}_n\cdot\mathbf{s}_\beta) - \left(\frac{\mu_\alpha}{\mu_\beta}\right)(\mathbf{s}_\beta\cdot\mathbf{r}_n)(\dot{\mathbf{r}}_n\cdot\mathbf{s}_\alpha)\right\}. \qquad (13)$$

Both angular momenta obey standard commutation rules (except $i \to -i$, since the indices are identified with axes in the body-fixed frame).

The angular momentum defined in eq. (13) can now be identified as the *vortex spin*; it is an essential element in both the symmetry approach, and the microscopic approach, to collective motion.

6. Generalizations of this symmetry approach

Although the $SL(3,\mathbb{R})$ model does indeed achieve a description of the nuclear rotational motion, and includes the vortex spin implicitly, it is desirable to include further collective degrees of freedom. To enlarge the scope of the model one now introduces the collective quadrupole degrees of freedom explicitly, a step which—although not foreseen at the time—is actually the key to further understanding of the vortex spin.

Here one runs into an interesting, purely group theoretic, difficulty: one *cannot* adjoin the five quadrupolar variables $Q_{ij}(\text{tr } Q = 0)$. The reason is that $\{\mathbf{J},\mathbf{T}\}$ must act on Q_{ij} but this cannot be, since there exists *no* five dimensional (non-unitary) irrep for $\overline{SL(3,\mathbb{R})}$. We must necessarily introduce the monopolar (tr $Q \neq 0$) degree of freedom.

This results in the *group of collective motions in three space*[5] denoted $CM(3)$, or more descriptively, $T_6 \textcircled{s} \overline{SL(3,\mathbb{R})}$ with the algebra having the commutation relations:

$$\left.\begin{array}{l}[J,J] \subset J \\ [J,T] \subset T \\ [T,T] \subset J\end{array}\right\} \quad J,T \text{ generate the } \overline{SL(3,I\!R)} \text{ algebra,}$$

$$\left.\begin{array}{l}[J,Q] \subset Q \\ [T,Q] \subset Q\end{array}\right\} \quad Q \text{ belongs to the non-unitary irrep [200],}$$

$$[Q,Q] = 0\} \quad \text{the six } Q_{ij} \text{ commute and generate } T_6.$$

How does this collective motion group, $CM(3)$, accord with the microscopic approach? We see that $CM(3)$ incorporates the three rotational degrees of freedom, three vortex spin degrees of freedom and two of the three (μ_α) deformational (or vibrational) degrees of freedom [since $CM(3)$ preserves volume, one degree of freedom is eliminated]. Thus we see that this algebraic (symmetry) approach to collective motion incorporates all but one of the collective degrees of freedom and three internal degrees of freedom of the microscopic approach. (The three center-of-mass degrees of freedom are easily incorporated and require no discussion.)

The $CM(3)$ model is sufficiently general that it includes the Bohr-Mottelson model of $\beta - \gamma$ vibrations as a special case, and indeed contains the generalization of this model to large deformations. To obtain this (Bohr-Mottelson) special case one need only remove all vortex motions (using the constraint $\mathcal{L}_{\alpha\beta} \to 0$) obtaining thereby irrotational flow (and liquid moments of inertia). One can recover the five quadrupole boson operators (and their conjugates) of the Bohr-Mottelson approach by a Wigner-Inönu contraction of $CM(3)$.

The collective motion group $CM(3)$ is analogous in its structure to the familiar Poincaré group, $\mathcal{P}$. Both groups have two invariants,[5] with the *mass* and *spin* invariants of $\mathcal{P}$ corresponding to the *volume* (Λ) and *vortex spin* (v) invariants of $CM(3)$. The analog of the rest frame for a Poincaré group representation is now the frame in which the mass-quadrupole tensor Q_{ij} becomes *spherical*, the deformation generators of $CM(3)$ being the analog of the boost operators. The angular momentum in the spherical frame is now the *vortex spin*, just as the angular momentum in the rest frame for the Poincaré group defines the intrinsic spin.

The vortex spin is the key to understanding the observed properties of rotational nuclei. In the absence of vortex spin one obtains irrotational collective motion, small moments of inertia, and spectra typical of vibrational degrees of freedom. Rigid motion (for which we have classically the constraint that the internal velocities are zero) corresponds to non-vanishing vortex spin and large moments of inertia. As the kinetic energy operator in eq. (11) shows, vortex spin couples the internal and collective degrees of freedom, but the question as to which vortex spins occur, and their mixing, is fundamentally a dynamical problem. It is interesting to note that the analogous problem for the Poincaré group—that is, embedding $\mathcal{P}$ in a larger structure combining irreps with different masses and spins—led to the introduction of supersymmetry. For collective motion, this same problem led to the introduction of the $Sp(6, I\!R)$ model.[9]

The $Sp(6, I\!R)$ model adjoins to $CM(3)$ the one remaining collective degree of freedom, by dropping the constant volume constraint and adjoining the generator of dilations. (This step by itself leads to a slight generalization of $CM(3)$ to $T_6 \textcircled{s} GL(3, I\!R)$.) In addition one adjoins six operators corresponding to a symmetric tensor made from the particle momenta:

$$p_{ij} = \sum_n p_{in} p_{jn}.$$

The resulting structure contains 21 generators and is quite remarkable for the large number of sub-models it includes:

(1) the $CM(3)$ model is a sub-group, so that irreps of $Sp(6, I\!R)$ will combine (Λ, v) irreps with different vortex spins. Since $CM(3)$ contains the Bohr-Mottelson model and the rigid rotor models as contraction limits, so also does the $Sp(6, I\!R)$ model;

(2) Elliot's $SU(3)$ model is a sub-group;

(3) the harmonic oscillator Hamiltonian is contained as a one-element subgroup so that $Sp(6, I\!R)$ includes the (orbital) shell-model.

This is important as a matter of principle since $Sp(6, I\!R)$ *per se* makes no reference to any internal degrees of freedom except the vortex spin (which is symmetric in the particle labels). By considering orbital shell-model states of the individual particles one can build up anti-symmetric states (adjoining spin and isospin) and generate collective $Sp(6, I\!R)$ states *properly obeying the Pauli principle*[10];

(4) there is a Wigner-Inönu contraction limit of $Sp(6, I\!R)$ which yields all the boson operators for $L = 0$ and $L = 2$ bosons. In this limited sense the model includes the Interacting Boson Model (IBM). However, these S and D bosons correspond to giant multipole degrees of freedom and hence are too energetic to be identified with the IBM. The $Sp(6, I\!R)$ model is unusual among spectrum generating symmetry models in that it contains the harmonic oscillator Hamiltonian as a (dynamical) generator. For a purely *kinematic* model, such as $T_6 \bigotimes GL(3, I\!R)$ (which also contains S and D bosons as a contraction), this (dynamical) criticism would not apply.

The $Sp(6, I\!R)$ model has been quite successful in accounting for experimentally observed nuclear structure.[9]

Closed shell nuclei typically belong to $Sp(6, I\!R)$ irreps vortex spin zero, implying irrotational collective motion, as observed. Rotational nuclei by contrast exhibit $SU(3)$ spectra, as in the Elliot model, but one finds that *the lowest $SU(3)$ states in the band correspond to angular momenta which are vortex spins*. Low lying, well developed rotational spectra are accordingly primarily due to vortex motions. This is an interesting re-interpretation of the meaning of the Elliot model.

The $Sp(6, I\!R)$ model is related to the orbital shell model, and derives much of its strength from this fact. But the Mayer-Jensen shell model involves spin in an essential way, and this is neglected (dynamically) in the $Sp(6, I\!R)$ model. Is it possible to include spin?

One of the basic features of the $CM(3)$ model is the fact that *vortex spin can be half-integer as well as integer*. In fact, the introduction of the $CM(3)$ model emphasized the conceptual advantage of half-integer vortex spin over the earlier particle-plus-rotator models that employed an unphysical distinction between nucleons in achieving half-integer angular momenta. Let us remark that half-integer vortex spin $(1/2, 3/2, \cdots)$ might offer conceptual support for symmetry group models[11] based on *pseudo-spin* in certain regions of the Mayer-Jensen shell model.

It is important open problem as to whether or not a larger symmetry group exists which contains half-integer $CM(3)$ irreps. It is easily proved that the $Sp(6, {\rm I\!R})$ will *not* work. (Since $Sp(6, {\rm I\!R})$ contains both $SU(3)$ and $SL(3, {\rm I\!R})$—with the two subgroups having $SO(3)$ in common—it is clear that only integer vortex spins can arise ($SU(3)$, unlike $\overline{SL(3, {\rm I\!R})}$, cannot have $SU(2)$ in the quadrupolar decomposition). If such a larger symmetry did exist, one might hope to extend the success of $Sp(6, {\rm I\!R})$ for the orbital shell model to the Mayer-Jensen shell model itself.

7. *Concluding remarks*

We have sketched above the story as to how Murray Gell-Mann contributed in an essential way to the theory of nuclear rotational motion, and have indicated briefly how these contributions still shape the most recent research in the field.

We would like to thank the organizers of the Gell-Mann Symposium for the opportunity of making these remarks, and we offer our best wishes to Murray on this happy occasion.

References:

1. R. F. Dashen and M. Gell-Mann, *Phys. Lett.* **17** (1965) 275; Y. Dothan, M. Gell-Mann and Y. Ne'eman, *Phys. Lett.* **17** (1965) 148

2. L. Weaver and L. C. Biedenharn, *Phys. Lett.* **32B** (1970) 326, ibid *Nucl. Phys.* **A185** (1972) 1

3. A. Bohr and B. R. Mottelson, *Nuclear Structure*, Vol. II, W. H. Benjamin, Reading, Massachusetts, 1975

4. L. C. Biedenharn, R. Y. Cusson, M. Y. Han and O. L. Weaver, *Phys. Lett.* **42B** (1972) 257

5. L. Weaver, R. Y. Cusson and L. C. Biedenharn, *Ann. Phys.* **102** (1976) 493

6. A. Y. Dzublik, V. I. Ovcharenko, A. I. Steshneko and G. V. Filippov, *Yad. Fiz* **15** (1972) 869; *Sov. J. Nucl. Phys.* **15** (1972) 487

7. W. Zickendraht, *J. Math. Phys.* **12** (1971) 1663

8. B. Buck, L. C. Biedenharn and R. Y. Cusson, *Nucl. Phys.* **A317** (1979) 205

9. D. J. Rowe, *Rep. Prog. Phys.* **48** (1985) 1419

10. V. Vanagas, *Group Theoretical Methods in Physics*, Proc. Intern. Seminar, Zvenigorod, 1982, Eds. M. A. Markov, V. I. Man'ko, A. E. Shabad, Gordon and Breach, Vol. A , N.Y. (1984) 259

11. J. N. Ginocchio, *Ann. of Phys*, **126** (1980) 234

Uncomputability, Intractibility and the Efficiency of Heat Engines[*]

Seth Lloyd

California Institute of Technology, Pasadena, CA 91125

Abstract

We consider heat engines that take both energy and information from their environment. To operate in the most efficient fashion, such engines must compress the information that they take in to its most concise form. But the most concise form to which a piece of information can be compressed is an uncomputable function of that information. Hence there is no way for such an engine systematically to achieve its maximum efficiency.

---

[*] This work supported in part by the U.S. Department of Energy under Contract No. DE-AC0381-ER40050

Heat engines take heat from their environment and turn it into work. We consider here engines that gather both heat and information and turn them into work. An example of such an engine is the Szilard engine [1], a one-molecule heat engine that turns information into work. Practical examples include engines that run off of fluctuations, and car engines that use microprocessors to achieve greater efficiency.

For an ordinary heat engine a Carnot cycle can in principle be carried out reversibly. Following a suggestion of Zurek [2], we show that engines that process both heat and information cannot attain the Carnot efficiency even in principle. We prove that to operate at the maximum efficiency over a cycle, such an engine must reversibly compress the information that it has acquired to its most compact form. But Gödel's theorem implies that the most compact form to which a given piece of information can be compressed is an uncomputable function of that information. Accordingly, there is no systematic way for an engine to achieve its maximum efficiency. Engines that get and process information are fundamentally inefficient: they suffer from *logical friction*.

Physics of information processing

The minimum amounts of dissipation required in principle by the operations of acquiring, registering, processing and erasing information have been extensively investigated. The following facts are well established [3-16]:

(1) It is possible to get information, and to store it in a register previously prepared in a standard state without increasing entropy [3-9, 16].

(2) It is possible to transform register states according to reversible logical operations without increasing entropy. All conventional logical operations can be embedded in reversible operations [7-8, 11-16].

(3) If a copy of a piece of information is accessible, the information can be erased without increasing entropy. If no such copy is accessible it is not possible to erase the information without increasing entropy. [3-10, 16].

The sort of information processing devices in use today generate much more than the minimum amounts of entropy required by these three facts. However, some logical devices, such as coupled Josephson junctions, are capable of coming quite close to these limits [16]. Our purpose is to put an absolute lower bound on the amount of dissipation required over a Carnot cycle. Our method is to apply the three facts above to an engine that gets, processes, and erases information, to show that such an engine cannot attain its maximum efficiency in a systematic fashion.

A heat and information engine

Consider an engine with the following properties. The engine can take heat from its environment and gather information. It has internal reservoirs and registers to store that heat and information. It possesses mechanisms that allow it to process both to do work, and to dispose of waste heat and information.

In the course of a Carnot cycle, the engine begins with all internal reservoirs and registers in standard states. It then takes in heat and information, stores heat in reservoirs and information in registers, and processes heat and information to perform work. Finally, it completes the cycle by restoring all reservoirs and registers to their standard states, disposing of any remaining heat and information as exhaust. In principle, each stage of the cycle can be carried out reversibly, without dissipation. To achieve its maximum efficiency, the engine must waste as little heat and information as possible.

Concentrate on the processes that involve information. First, the engine gets information about its environment, reducing the environment's entropy by $\Delta S = -\sum_i p_i log_2 p_i$, where p_i is the probability of the ith state that the engine can distinguish, and where we take the logarithm to the base 2 to make the connection between statistical mechanics and information theory. In accordance with fact (1) above, we assume that this process takes place without dissipation.

The engine then records in its register a message of length l_i, indicating that the environment was measured to be in its ith state. A message is simply a particular pattern of a subset of the degrees of freedom that make up the engine's register. For the sake of convenience, we assume these degrees of freedom to be discrete and binary. The length of a message is equal to the number of degrees of freedom fixed in registering the message. Once again, the registering process can be carried out without dissipation in principle.

The engine now has information about the environment in its register, in the form of a message of length l_i. The engine manipulates and transforms this information systematically, according to logical rules, and uses it to do work. By fact (2), this manipulation can be carried out without dissipation as long as information is transformed according to logically reversible operations. After all work has been done, the engine is left with a message of length l'_i. All remaining degrees of freedom in the register are in their standard states.

To complete the Carnot cycle, the engine must restore each degree of freedom taken up by the final message to its standard state—it must erase any information left in its register. But to erase one bit of information requires a unit increase in entropy. In restoring the register degrees of freedom to their standard states, the engine generates entropy $\sum_i p_i l'_i$ on average.

Note that the overall entropy actually *decreases* during the first stage of the cycle. We now prove, however, that entropy is non-decreasing over the complete cycle. (The second law of thermodynamics, as usually stated, holds for cyclic processes only. The normal form of the second law must be modified if it is to hold for engines that do not work in a cyclic fashion, simply accruing information without erasing it [2,10].)

Theorem 1. *The total change in entropy over one cycle of an information/heat engine is*

$$\sum_i p_i l'_i - (-\sum_i p_i \log_2 p_i) \geq 0.$$

Proof: The total change in entropy over the cycle is the sum of the increase in entropy due to erasure, $\sum_i p_i l_i$, and the decrease in entropy as information is acquired, $-(-\sum_i p_i \log_2 p_i)$. We need to show that this sum is ≥ 0.

To insure that the engine is able to distinguish between the various messages, each message must include a specification of which degrees of freedom are fixed in the pattern that makes up the message. In the language of coding theory, a set of such messages forms a prefix-free code [17]. The length of the different final messages then obeys the Kraft inequality: $\sum_i 2^{-l'_i} \leq 1$. Given a set of non-negative numbers a_i such that $\sum_i a_i \leq 1$, we have $-\sum_i p_i \log_2 a_i \geq -\sum_i p_i \log_2 p_i$. Let $a_i = 2^{-l'_i}$. Theorem 1 follows immediately.

In order for the engine to attain its maximum efficiency, two requirements must be met. First, the minimum message length for the ith message must coincide with $-\log_2 p_i$. Second, the engine must be able to compress the ith message to this minimum length. The first requirement can be met, modulo the difficulty of trying to match continuous $-\log_2 p_i$ with discrete l'_i. In Shannon-Fano coding [17] the messages have length $l'_i = \text{Int}(-\log_2 p_i)$, where Int gives the next integer greater than its argument. Such a coding gives $\sum_i p_i l'_i = -\sum_i p_i \log_2 p_i + O(1)$, where $O(1) \leq 1$ (Shannon-Fano coding is not necessarily the optimal code: Huffman coding is even better, but is harder to construct [17]).

A way of assigning message lengths that attains the maximum efficiency exists, but that does not mean that an engine can actually obtain such a coding through reversible logical operations. We prove below that the second requirement for maximum efficiency cannot be met in general. If the engine is pre-programmed with the probabilities for all possible environmental configurations—if its message lengths are precisely tuned to the frequencies of its inputs in advance—then it can operate without dissipation. But if probabilities are not given *a priori*, then the best the engine can do is try to estimate the frequencies of its inputs, try to tune its message lengths to the $-\log_2 p_i$, and in the meanwhile, try to compress each message to the shortest length possible.

The goal of maximum compression is unattainable in principle, however, and not simply because of normal friction, which can be done away with in principle in the limit of infinitely slow operation. Rather, there are deep logical reasons that prevent an engine from attaining its maximum efficiency in a systematic fashion. Theorem 1 implies that to attain its maximum efficiency, the engine must compress the information in its register to its most compact form before the end of the cycle. But the most compact form to which a given piece of information can be compressed is in general an uncomputable function of that information [18-22]. Therefore the engine can attain neither the maximum compression of information nor the maximum efficiency using recursive operations. An optimal way of coding information exists, and can be attained by pre-programming or by accident (the message lengths assigned just happen to be the optimal ones), but in the absence of good luck and outside guidance there is no systematic way for the engine to achieve its maximum efficiency.

Gödel's theorem and Berry's paradox

The uncomputability of the most compact form for a given piece of information rises from a variation of Gödel's theorem. Gödel's theorem says that in any formal system there are true statements that cannot be proved [23]. The theorem is based on the Cretan liar paradox (Epimenides the Cretan once said "All Cretans are liars"). Gödel's contribution was to include the analogous statement, "This statement cannot be proved," in a formal system. If the statement is provable, then it is false, which is impossible. Therefore the statement is true, but cannot be proved.

The most compact form to which a given piece of information can be compressed exists, but the proof of Gödel's theorem can be modified to show that it is an uncomputable function of that information. This result can be derived from a formalization of Berry's paradox, first proposed by Russell [24]: "Find the smallest positive integer that requires more characters to specify than there are characters in this sentence." This sentence seems to specify an integer, and yet cannot.

Chaitin included this paradox in a formal system of computation as follows: suppose that the length of the shortest program to specify a particular number is a computable function of that number. Then given the basic axioms of a theory, of length n, it is possible to write a program of length $n + O(1)$ that goes through all numbers in order until it finds the first number that requires more than $n + O(1)$ bits to specify. "The program finds the first number that it can be proved incapable of finding." The only resolution of this paradox is that the shortest program to specify a particular number is not a computable function of that number. Applied to our information/heat engines, this result implies that the shortest message to which a given message can be compressed is not a computable function of that message. We now give a formal proof to this notion.

Logical limits on the efficiency of heat engines

We assume a heat/information engine as in theorem 1. In the most general case, the engine is pre-programmed with partial knowledge of the probabilities of different events that it is to register. The engine has some *a priori* knowledge, but that knowledge is incomplete. We then have

Theorem 2. *Given an engine pre-programmed with n bits of information. If the minimum length to which any of the messages in the engine's register can be compressed is $\geq n + O(1)$, then the engine cannot attain its maximum efficiency using recursive operations.*

Proof: Theorem 1 implies that to attain its maximum efficiency, the engine must compress the information in its register to the message of the minimum length. When the engine first gets information it registers it as a message m. Consider the set of messages $M(m) = \{m_k = F_k(m) \text{ for all } k\}$, where $\{F_k\}$ is the set of reversible transformations that the engine is capable of effecting, given the information pre-programmed into the engine. We are interested in the case in which the F_k are recursive functions. The operation of the engine is then equivalent to a formal system with n bits of axioms.

It is a theorem due to Chaitin that in a formal system with n bits of axioms it is impossible to prove that the minimum length to which a particular message can be compressed is greater than $n + O(1)$. Consider the subset of $M(m)$,

$$M_{\min}(m) = \{m_{k'} : l_{k'} \leq l_k \text{ for all } k\},$$

where l_k is the length of m_k. The minimum length to which m can be compressed is then

$$l_{\min}(m) \equiv l_k : m_k \epsilon M_{\min}(m).$$

Chaitin's theorem implies that for m' such that $l_{\min}(m') \geq n$, none of the members of $M_{\min}(m')$ is a recursive function of m. If the engine has such an m' in its register at any point, it cannot compress it to its minimum length using recursive operations, and cannot attain the maximum efficiency in a systematic fashion. Q.E.D.

In the special case $n = 0$, Theorem 2 immediately implies the following:

Corollary. *An engine given no a priori information cannot attain its maximum efficiency in a systematic fashion.*

Theorem 2 implies that an engine with no *a priori* information can compress no messages whatsoever to their minimum length. Note that even with pre-programming and successful message compression, an engine may be operating very inefficiently if the lengths of the compressed messages are not closely related to the probabilities of the events that they register. The algorithm that the engine follows must reflect the order of the information that it is processing.

Discussion

Another way of stating our result is that a heat engine is logically incapable of exploiting all the order in its surroundings. The practical counterpart to this result has been clear for some time. In real heat engines, logical friction—the tiny amount of dissipation required by excess erasure—is dwarfed by mechanical and electrical friction. Nonetheless, the prescription for efficient operation implied by theorems 1 and 2 holds good. Theorem 1 requires that to operate efficiently the engine should be jealous of memory space, and should tune the length of its messages to the probabilities of the events they signify. Theorem 2 implies that perfect tuning of message lengths is impossible, but that maximal efficiency can be approached by pre-programming the engine to take advantage of probable order in its environment. In addition, the practical experience of the ages suggests that needless mechanical dissipation be avoided.

Finally, applied to human beings, logical friction implies that the ultimate scientific theory—the most concise way of expressing the information that we have about the universe—is unattainable in principle [25-26].

References

1. L. Szilard, *Z. Phys.* **53** (1929), 840.

2. W. H. Zurek, *Nature* **341** (1988), 119.

3. R. Landauer, *IBM J. Res. Dev.* **5** (1961), 183.

4. R. W. Keyes, R. Landauer, *IBM J. Res. Dev.* **14** (1970), 152.

5. R. Landauer, *Phys. Rev. Lett.* **53** (1984), 1205.

6. R. Landauer, *Phys. Scri.* **35** (1988), 88.

7. C. H. Bennett, *IBM J. Res. Dev.* **17** (1973), 525.

8. C. H. Bennett, *Int. J. Theor. Phys.* **21** (1982), 905.

9. C. H. Bennett, *Sci. Am.* **257** (1987) 108.

10. S. Lloyd, *Phys. Rev.***A**, to be published.

11. E. Fredkin and T. Toffoli *Int. J. Theor. Phys.* **21** (1982),1219.

12. P. A. Benioff, *Phys. Rev. Lett.* **48** (1982), 1581.

13. P. A. Benioff, *Int. J. Theor. Phys.* **21** (1982), 177.

14. R. Feynman, *Int. J. Theor. Phys.* **21** (1982), 467.

15. D. Deutsch, *Proc. R. Soc. London* **A 400** (1985), 145.

16. K. K. Likharev, *Int. J. Theor. Phys.* **21** (1982), 311.

17. R. W. Hamming, "Coding and Information Theory," 2nd edition. Prentice-Hall, Englewood Cliffs, New Jersey, 1986.

18. G. J. Chaitin, *Sci. Am.* **232** (1975) 47-52, reprinted in "Information, Randomness & Incompleteness," World Scientific, Singapore, 1987.

19. G. J. Chaitin, *IEEE Transactions on Information Theory* **20** (1974), 10-15, reprinted in "Information, Randomness & Incompleteness," World Scientific, Singapore, 1987.

20. G. J. Chaitin, *IBM J. Res. Dev.* **21** (1977), 350-359, reprinted in "Information, Randomness & Incompleteness," World Scientific, Singapore, 1987.

21. G. J. Chaitin, *Int. J. Theor. Phys.* **22** (1982), 941-954, reprinted in "Information, Randomness & Incompleteness," World Scientific, Singapore, 1987.

22. A. K. Zvonkin and L. A. Levin, *Uspekhi Mat. Nauk* **25(6)** (1970) 83-124.

23. K. Gödel, *Monatshefte für Math. und Phys.* **38** (1931), 173-198, reprinted in "From Frege to Gödel," ed. J. van Heijenoort, Harvard University Press, Cambridge, Mass., 1967, and in "The Undecidable," ed. M. Davis, Raven Press, Hewlett, N.Y., 1965.

24. B. Russell, *Am. J. Math.* **30** (1908), 222-262, reprinted in "From Frege to Gödel," ed. J. van Heijenoort, Harvard University Press, Cambridge, Mass., 1967.

25. G. J. Chaitin, "Information, Randomness & Incompleteness," World Scientific, Singapore, 1987.

26. R. J. Solomonoff, *Inform. and Contr.* **7** (1964) 1-22, 224-254.

The New Mathematical Physics
by
I.M. Singer

It is an honor to have been selected among Murray's many friends and colleagues to speak to you this afternoon. No doubt I have been chosen because of the mathematical component present in high-energy theory today. Before concentrating on the interface of elementary particle physics and modern geometry, I'd like to record my own pleasure in knowing Murray this past decade. Perhaps it's just as well we didn't get acquainted earlier; I think he would have frightened me to death. You all are aware of Murray's great intellectual powers; but to me, equally amazing, is his enthusiasm for all creative endeavors, large and small. More than anyone, he firmly believes that the human mind and the human spirit can cure the ills of society. This birthday celebration expresses his personality in several ways. The diversity of topics reflects his many interests. And the theme stresses his positive view of the future.

Our charge was to pick some subject — mine is mathematics and physics — and discuss its present status and future prospects. Like twin stars, the two subjects have influenced each other greatly over the centuries, sometimes overlapping significantly, sometimes going their separate ways. In the fifties and sixties there was little contact — perhaps even some hostility. Physicists believed that too much mathematics hindered physical insight; some older ones still do. Mathematicians required more mathematical precision than physics deemed necessary and were developing abstract structures for their own sake. Some still feel mathematics is "pure" and not to be tainted by physics. [Workers in constructive and algebraic quantum field theory were left to their own devices; we are only now beginning to appreciate their important insights.] What a pity! History teaches us how fruitful the overlap can be — witness the great developments: classical mechanics and the calculus, relativity and differential geometry, quantum mechanics and operators on a Hilbert space, symmetries of nature and group theory.

The recent interaction has been very fruitful. It began for me in 1975 with the Wu–Yang dictionary showing, among other things, that vector potentials in non-abelian gauge theories are the same as connections on a fiber bundle. By 1977 mathematicians understood that instantons were self dual solutions to the Yang–Mills equations of motion and could classify them; [for the non-expert, these are special nonlinear generalizations of Maxwell's equations]. In the early eighties, Witten introduced supersymmetry into mathematics illuminating both Morse theory and index theory. I date the rebirth of string theory with the Green–Schwarz work on anomalies, which has a large mathematical component. About the same time the Polyakov measure and two-dimensional conformal field theories were connected to line bundles on moduli space. The current excitements are topological quantum field theories and the expression of topological invariants in quantum mechanical terms.

These developments affected the amount and variety of mathematics young physicists need to know: characteristic classes, the cohomology ring of a Calabi–Yau manifold, Lie groups — their representations and homotopy groups, affine Lie algebras, diffeomorphism groups, index theory for geometric operators, moduli spaces for Riemann surfaces, modular tensor categories. It is amazing to me how quickly young physicists working in the modern interface of geometry and elementary particle physics have mastered the basics of the subjects I have listed. Mathematicians, except for a few experts, have been much slower to learn important aspects of quantum field theory: perturbation theory, renormalization and the renormalization group, supersymmetry in path integrals — subjects we need to master if we are to become more than consultants on physical problems. Young physicists require more mathematics than what they learn in graduate physics courses. And mathematicians need to be reminded of E. Cartan, Riemann, and Poincaré in order to temper their concentration on rigor.

The new ideas from physics have had great impact on mathematics. Self dual Yang–Mills fields (instantons) are the key ingredient in Donaldson's work exhibiting four-manifolds that are not smoothable. The combined work of Donaldson and Freedman imply the existence of exotic four-dimensional space times much different from what our intuition provides. Supersymmetry has illuminated index formulas, although the proofs remain difficult. New invariants for knots arise from Witten's interpretation of Chern–Simons quantum field theory as a topological theory closely related to conformal field theory. Supersymmetryhas added a new element to classical Riemann surface theory in the form of super-Riemann surfaces.

It must be said that the developments on this abstract side of high energy physics which we have been discussing, string theory et al., have not led to experimental predictions or to the solutions of longstanding problems in QCD. But within the contexts of the subject, mathematical techniques have been essential. String theory models begin with the representation theory of Kac–Moody groups. They need Calabi–Yau metrics, the more explicit the better. Anomalies are best understood in terms of the families index theorem. The geometry of moduli spaces for Riemann surfaces is at the base of two-dimensional gravity theories.

The contact between modern geometry and elementary particles continues unabated and with increasing vigor. What can we expect in the future? I think the liaison will flourish for a long period of time. Although mathematics and physics have different goals, they share many themes. Symmetry is one of them. It clarifies and simplifies much in physics (the eight-fold way, for example — an exquisite art form in Murray's hands). Today we are extending our studies of finitely parameterized groups to infinite groups (Virasoro, loop groups) using our past experience of the representation theory of Lie groups. We seek a deeper symmetry in string field theory that may suggest how to treat non-perturbative effects and how to make predictions in lower energy regimes.

Geometry is a second common feature. Gauge theories and general relativity are geometric theories. Moreover, the laws of physics are independent of coordinate systems. Global geometry and topology study invariants independent of local systems. One could also say that quantum field theory is quantum mechanics over infinite dimensional geometries.

With experimental data harder and harder to come by, high energy theorists must rely more heavily on the heritage of physics for guidance and in that heritage, geometry and symmetry loom large. In my view not only will the constructive interaction be sustained, but in fact, there is a new branch of mathematical physics developing — an abstract theoretical physics — which brings more sophisticated mathematical methods to bear on problems of physical origin.

Perhaps the immediate agenda is a deeper understanding of three- and four-dimensional space times. Although I have emphasized the overlapping areas between mathematics and physics, I should also mention some differences. They have different aims — the study of natural phenomena vs. the study of "coherent systems" of which nature is only one. They have different time scales. There is a sense of immediacy in physics that mathematics does not share.

It took almost two thousand years to understand two-dimensional geometry, from Euclid's parallel postulate to the discovery of hyperbolic geometry by Bolyai, Gauss, and Lobachevsky in the early nineteenth century. In these geometries the sum of the interior angles of a triangle is not one hundred and eighty degrees. Gauss went surveying, measuring angles from mountain tops, to see whether the space we live in is indeed flat, i.e., Euclidean. He was testing the nature of space by classical means. Today we could say we are examining three- and four-dimensional spaces by quantum mechanical means. Inspired by Atiyah, Witten recently began finding topological invariants for manifolds by path integrals, i.e., appropriate quantum field theories. Both mathematicians and physicists are exploring the potentialities and limitations of these new methods. These developments together with the earlier works of Donaldson, Freedman, and Thurston make it safe to predict that we soon will have a clear picture of three- and four- dimensional spaces.

Inevitably these new insights will have some impact in physics. Both our conception of the big bang and the expanding/contracting universe are based on our presently limited understanding of four-dimensional geometry. New exotic space times and new geometries must be considered in our attempts to understand the "beginning" and possible "ends" of our universe.

In mathematics, there is a faith that what one finds interesting will ultimately be useful. That faith has been justified. There is a similar belief developing among young theorists in physics. The new liaison has had many applications to mathematics, as I have indicated. There is some hope that the insights of the new abstract theoretical physics will ultimately help solve nonperturbative problems in quantum field theory.

"Is Quantum Mechanics for the Birds?"
V.L. Telegdi
ETHZ, Swiss Federal Institute of Technology
Zurich, Switzerland

I would like to correct a misrepresentation made by several of the preceding speakers. We are not celebrating the sixtieth anniversary of Murray's birthday. We are celebrating the sixtieth anniversary of his conception. His actual birthday is in September.

We are, of course, celebrating Murray Gell-Mann, whom I've known now since 1951. We joined the University of Chicago the same year, a few months apart. And before I go into the more scientific part of my lecture, I think you might be interested in the origin of the name "Murray." Presumably it was, like some other first names, derived from a surname. And these, in turn, often come from geographical names, in the present case, from a Scottish province, "Muraih." Already in 1203 we find a William de Moravia, and in 1317 an Orland de Morris, and in 1327, an Andrew Muraih. [This does not prove the point, because these family names could well have come from "Murie," the Middle English form of Merry.] As a first name, it has also been surmised, as I see from a book by Partridge, *Name This Child*, that "Murray" comes from "Murrey" a word for dark red or eggplant colored, an adjective which in turn presumably comes from mulberry, in turn connected to maroon. Which brings us back to Murray's favorite color of corduroy jackets at the University of Chicago.

Now having explained the word Murray, I cannot refrain from giving you the origin of the name Gell-Mann. It is true his family came from Austria, but from the Austria of the pre-1918 days when the empire extended all the way to the Russian border. And this Gell-Mann is presumably a russification of "Hellmann," which in turn is almost demonstrably a corruption of "Heilmann" or "healer," a profession often practiced by minorities in the empire.[*] I cannot tell you how the hyphen got

---

[*] Experts in onomastics assure us that Hellmann is not derived from Heilmann. Nor is it likely that a Scottish name in "de," like William de Moravia, would come from an adjective meaning

slipped in, but as consolation award I can inform you that the word "hyphen" has nothing mystic about it: It's from the Greek "ὑφέν" for "under" and "one," so it's a thing that comes under one category. That explains the word.

So now, since you've heard all these very very learned people speak, you may think that my talk here is for the birds. To please Murray it is indeed a little bit about the birds, or rather, about one specific bird, the bearded vulture. This animal, which exists in Switzerland, is called in Latin *Gypaetus barbatus aureus*, and it's a lovely, much-maligned creature. The first slide[†] shows the bearded vulture, which is known in the German part of Switzerland as *Bartgeier*, (just a German translation of its English name) as *Gypaète barbu* in the French part and as *Avvoltoio barbato* and *Tschess barbet* in the Italian and Romanche parts respectively. It is not directly relevant, but surely amusing, to also know its Spanish name, which is *Quebranta huesos*, which means "bone breaker."

Well, anyhow, I told you that this bird is much-maligned, and I'll show you a classic Swiss painting. Here you see this bird is carrying away a young sheep, but now the name *Lämmergeier*, has been abandoned, because apparently it's just a fable that he actually does that; as I said he was much-maligned. And now I'm really talking to Murray: The Eidgenössische Vogelwarte, which is the Federal bird-watching station in Sempach, has very recently given me the sensational news that this bird has lately been *eingebürgert*, a term which means naturalized, for humans an expensive and difficult procedure in Switzerland. But it here signifies that bird pairs were raised in captivity, released and left to reproduce themselves in their natural habitat. And that, Murray, is one more reason for you to visit us in Switzerland.

Why did I give the title "Is Quantum Mechanics for the Birds?" to this talk? I had the great privilege of being able to come repeatedly to Caltech over the past ten years, and I think I occupy a unique position here, thanks to my friends and to the whole administration, namely that of a permanent visiting professor. During these

---
"merry" rather than a place name. *ed.*
[†] For technical and copyright reasons, the photographs shown in these slides could not be reproduced here.

visits I always kept asking Murray about the foundations of quantum mechanics. I thought the time had come for him to make his contribution to that topic. We heard a wonderful talk by Jim Hartle about the progress he and Murray have made. My own discussions with Murray can, as far as I'm concerned, best be summarized by a German proverb, and that is, *"Ein Narr kann mehr fragen als sieben Weise antworten,"* which means that "A fool can raise more questions than seven wise men answer." And I am going to give you a list of the questions that were asked, and perhaps of some that were answered.

But before I come to that, I would like to bring up another point, one that the Provost of this institution has so eloquently stated, namely that Murray is perhaps the Renaissance man of our time. You know about his interest in wildlife, in nature preservation, and in social affairs. But he has many many other talents that you are probably entirely unaware of. These I would like to illustrate with the following slides: The first of these shows Murray directing a symphony. This was of course in his younger years, but already quite an impressive achievement. I think they were playing Brahms. Now that is something some would have suspected. But just watch the second one. Here he is seen just at the point of winning a fencing contest. The third shows Murray and another chef jointly preparing an elaborate dish. This is the slide that impressed my wife most. We knew he appreciated food, but we didn't know he could prepare it too. The fourth picture shows him as an equilibrist, standing on top of a small platform on the apex of a pyramid of bottles. A remarkable sense of symmetry and equilibrium! As a special treat I shall show the last of these slides only at the end of my talk. (Incidentally, glossy prints of these pictures are available starting next week for about ten dollars a piece; that is to defray my own expenses.)

Back to science. Now, to most of these questions I'll list, I'll not give the answers, perhaps address them a little bit in the conclusion, as you have heard most of them from Hartle already. The first question is the inclusion of the observer, the well-known collapse of the wave function, the phenomenon known as "Schrödinger's cat," the rôle of consciousness in observation. Next, the rôle of cataclysmic amplification in observing elementary and in particular atomic processes. In all the textbook experiments that you hear about, concerning isolated systems, electrons and atoms, there is always cataclysmic amplification between you and the process to be observed. Then, as Hartle already asked, "What is a classical object?" "How small is small enough?" In other words, is something made of a hundred atoms classical? Of ten? Of a million? We don't know. Then, a phenomenon often discussed, "non-locality," the famous EPR business. Another question concerns the wave function of the Universe. With whom does the Universe interact? When you teach or learn elementary quantum mechanics, predictions are always supposed to be statistical, but there's only one Universe, so how are you going to apply what you learned in school? Thus the question is, how can you interpret the Universe statistically? What happens if there are other Universes? A more technical question: is there meaning to ordinary quantum mechanics outside of a complete field theory? One last question: We've all heard about black holes. They swallow things. What happens to the wave function that was swallowed by a black hole?

These are the kind of problems that I had the privilege of discussing with Murray. Anticipating the conclusion that I'm supposed to give, I think that high-energy physics will *not* yield the exciting results that Professor Fritzsch mentioned, because in the foreseeable future the jump in energy will be too small. On the other hand, I think that the textbooks of the next thirty years will present quantum mechanics very differently from the way it is taught now, and it certainly will be a quantum mechanics that includes the Universe in which we live, and it certainly will be one that is due to the efforts of people like Gell-Mann and Hartle.

Now there is one phenomenon in quantum mechanics which I think has not received much attention and which is simple, and that is the quantum mechanics of non–observation. Namely, what happens if you don't do something, or don't see something? Like the old–fashioned way of saying, "If you didn't hear or see the tree fall in the forest, did it fall, or didn't it fall?" This is just a "classical" example. I'm going to talk about a situation concerning which an unbelievable amount of literature (not quoted here) has been accumulated. Nevertheless, when I asked during the coffee break several very distinguished physicists in this audience, including one Nobel

Prize winner, "Have you ever heard about the "shelved electron?" to my enormous satisfaction none of them answered in the affirmative. So that will be my example of a case of non-observation. Since I am an experimental physicist, I'm going to talk about an actual experimental situation.

The remarkable thing here is, that this is one of the few, almost unique, cases, where the experiment is really like the ones talked about in textbooks: You're dealing with a *single* atom. None of the phenomena that I'm going to describe could happen if you had ten or twenty of them at the same time. Current experimental technology enables you to store atoms for an indefinite amount of time in a small amount of space, to subject them to external influences and to see what happens. So one is very close to certain idealized textbook situations that technology did not allow you to handle before, and all of a sudden you are supposed to be able to describe it correctly. The "shelved" electron, I would like to underscore, was proposed[1] and later also experimentally investigated by Hans Dehmelt in Seattle, by the man who has done those transcendentally beautiful experiments on single stored electrons that everybody should know about*. Now, I'll take some sentences from the literature, which tell what people thought about this experiment when it was proposed. I'll read it here: "A single atom can illustrate to the naked eye the nature of quantum mechanical time evolution. Even if the practical spectroscopic applications were never realized, the pedagogical value of this experiment for the understanding of quantum mechanical measurements *cannot be overemphasized.*" Italics mine. "Like the Stern-Gerlach experiment, the Dehmelt scheme has far-reaching consequences, demonstrating in a way how quantum mechanical measurements work". I don't subscribe to the latter sentence, but it gives you a flavor.

So now, what is this "shelved electron" supposed to be? It's supposed to be the following situation (see Fig. 1a): You take an atom, such as actually exist. Not some fictitious atom, but a real atom. And let this atom have two excited levels – I apologize to the non-physicists, at this point you will lose all contact with me –

---

* The year of this talk, Dehmelt received the Nobel Prize for those experiments

one, E_b, that's strongly coupled to the ground state, G, and the other one, E_r that is very weakly coupled to it. The subscripts recall the wavelengths, respectively red and blue, of transitions from G. Dehmelt calls the level E_r the "shelf," for reasons which will become apparent in a minute. I'll have to introduce a certain amount of notation, e.g., the widths Γ_r, Γ_b of these levels. E_b of course is much broader than E_r. If you apply external laser fields to make these transitions, which you do, then you will have the so-called Rabi frequencies Ω_r, Ω_b, and if the frequencies that you apply to make the transitions are not exactly the differences between the levels and the ground state, you will have two detunings, Δ_r and Δ_b.

Now, what is supposed to happen? Assume you didn't have the level E_r here at all. It is very weakly coupled, but now we're going to consider zero coupling. Then when you shine in light of the appropriate color (blue), fluorescence will occur, and the same color photons that you sent in – blue – will come out. And with a modest laser you can have millions of them come out, because of the very strong coupling. Now, if you have two lasers, a blue and a red one, then the classical point of view, that of Dehmelt, says that in those rare (because we're dealing with a single atom!) cases when you manage to induce a transition onto the "shelf," the fluorescence will cease, because you can no longer make the "blue" transition. In other words, when the electron is sitting on the "shelf," one has to wait until it falls from it. Once it has fallen down, then you can kick it up again, and the fluorescence will resume.

This is a rather convincing elementary point of view. You see it here (Fig. 1b), a burst of fluorescence, another burst of fluorescence, in between there are periods of darkness; we indicate the mean lengths of these periods with T_D and T_B respectively. The purpose of any theory is to predict the ratio of these periods. Let me please point out that according to the semi–classical explanation which I gave, the fluorescence can commence only after a *red* photon has come out. Is that clear? It's then that you are back in the ground state. You can also say that every time you have a photon come out, the wave function collapses to that of the ground state.

Now, Hans Dehmelt is not only a great genius, (an epithet I do not apply to many people), but he is also a very practical man, and he wanted to turn this into a "have your cake and eat it too" spectroscopy. The dream of every spectroscopist is to deal with a very narrow line. The narrower the line, the better you can define its frequency. Of course, the narrower the line, the harder it is to make a direct fluorescence experiment. "So," says smart Dr. Dehmelt, "I'll do this bright/dark interruption as a function of the red frequency, and then I'll profit from the strong transition of the blue one, and just look when I get the most interruptions (dark periods)." So in this great scheme, the optimal ratio of the dark to the bright periods would occur when you are resonant with the red level, and it could have a width which is that of the red level, i.e., very very narrow. So this would be the "have your cake and eat it too" spectroscopy.

It's very interesting that this idea was not accepted. In fact, a conference of atomic physicists was held, I think in 1986, in Copenhagen, before any experiments were done, and half the experts said the telegraph pattern would occur like Dehmelt had said, and the other half said, "No, you people have forgotten about the principle of superposition in quantum mechanics. Just ignore this shelved stuff." They continued, "There will be a superposition of the states, and what will happen is that there will be many many blue photons and occasionally a red one. No gaps in between." Fifty percent of the people were of that opinion. I'm not here to defend those people. Maybe I should anticipate that Dehmelt was right. Not rigorously right, but essentially right.

I'm now going to give you the description of this phenomenon in the simplest possible terms. Among the many papers, I've chosen the only one that I think is not too mystic. But before I describe it let me make another comment: This Dehmelt phenomenon was hailed in the world of atomic physics as a great breakthrough: one could now, for the first time, "see a quantum jump microscopically." It's absolute nonsense of course, because already Rutherford saw quantum jumps through the scintillations of individual α-particles. What is new here, is that while you can't shoot back the same α particle into the same nucleus, here it is the same atom making transitions over and over again. But people keep waving copies of Schrödinger's 1935 papers at you, saying "Ah, but there *are* quantum jumps" – that is not related to the real issue here.

What are the issues? I shall report a discussion as written down practically during the aforementioned Copenhagen conference by two French physicists from the famed Ecole Normale: Claude Cohen-Tannoudji and Jacques Dalibard. It is published in Europhysics Letters[2] which already proves that it's a short paper. Their approach is a very simple one (for the physicists in this audience). They simply say we have to describe not the atom, but the *atom plus the laser field* in which it is bathed. Denoting the number of red (blue) photons in the laser cavity with $N_r(N_b)$, one deals with three relevant "dressed" states here: $|\varphi_1\rangle = |G, N_g, N_R\rangle$, coupled to $|\varphi_1\rangle = |E_b, N_b - 1\rangle$ and $|\varphi_2\rangle = |E_r N_b, N_r - 1\rangle$. You will readily recognize that except for detuning we're dealing with three quasi-degenerate states.

Now, the French are very good at notation, but I've outdone them. An even better notation would be this one, $|\varphi_0\rangle$, ground state, so many blue, so many red photons, in the cavity. And now there is a different category of photons, fluorescent ones, denoted by N'_r, N'_b, of which there are none in that particular state. What we have to compute are transitions from state without the fluorescent photons, to states with the fluorescent photons. You can write down the equations by inspection, (see Table 1) and then you have to compute the probability of *not* seeing a photon at the time τ, if you saw one at $t = 0$. It is so trivial to solve these equations that even I can do it. Square the amplitudes, and you will find that the probability for that to happen; there is a fast and slow component as you might well imagine, and generally there are three time constants because we're dealing with three equations.

Table 1 gives the solution of these equations. I shall first discuss the weak intensity ($\Omega_b \ll \Gamma_b$) case. The three time constants are here. The first one of course is essentially the lifetime of the blue state, and then there is another one, a long one. There are two short time constants. The long time constant (forget about the short ones which are sort of trivial, this is just the absorption rate from ground state to blue state) is of course the decay rate Γ_r, the spontaneous decay rate of the red state, as anybody might imagine, plus the stimulated decay rate from that state. So now then if you've computed that, you'll readily find the resonance. What resonance? The resonance is in the ratio of dark to bright periods T_D/T_B. Setting the blue laser on resonance ($\Delta_b = 0$), you will sweep the red laser. When you have tuned it in perfectly, you get a magnificent resonance, half the time it's bright, half the time it's dark (see Fig. 2[a]). Now you see quantum mechanics at work. It's not the natural widths of the states that determines the width of the resonance, but the widths including the induced transition. What determines that width are the parameters of blue transition, because the ground state G, strongly coupled to the excited state E_b, has been broadened by that transition, and so when the red photon goes down to the ground state, that is the width you get. Note that Ω_b^2/Γ_b is the blue pump rate. The observed width Γ_{obs} shown was derived assuming that the spontaneous and induced widths of E_r are made equal. Since $\Gamma_{obs} \gg \Gamma_r$, it would seem that the expected "have your cake and east it too spectroscopy" fails! No: it *can* be salvaged, by *alternately* exposing the atom to red (shelving) and blue (interrogating) laser light.

Table 1

Equations of motion:

$$i\dot{a}_0 = \frac{\Omega_b}{2}a_1 + \frac{\Omega_r}{2}a_2,$$

$$i\dot{a}_1 = \frac{\Omega_b}{2}a_0 - \left(\Delta_b + \frac{i\Gamma_b}{2}\right)a_1,$$

$$i\dot{a}_2 = \frac{\Omega_r}{2}a_0 - \left(\Delta_r + \frac{i\Gamma_r}{2}\right)a_2.$$

In the weak intensity limit ($\Omega_b \ll \Gamma_b$) the system has two short (1, 2) and a long time-constant (3):

$$1/\tau_1 = \Gamma_b/2, \ 1/\tau_2 = \Omega_b^2/2\Gamma_b \text{ and } 1/\tau_3 = \Gamma_r/2 + (\Gamma_b/2)(\Omega_r/\Omega_b)^2.$$

Assuming, in τ_3, the spontaneous and induced rates to be equal, one obtains

$$T_D/T_B = 1/[2 + (\Omega_b^2/2\Gamma_b)^2\Delta_r^2],$$

i.e., a Lorentzian of width $\Gamma_{obs} = \sqrt{2}\Omega_b^2/\Gamma_b$ peaking at 1/2.

Even more remarkable is what happens when you go to the strong intensity case. When, with a very strong wave, you saturate the blue transition, then you find that the states $|\varphi_i\rangle$, which were quasi-degenerate, split up and you get the magnificent picture given in Fig. 2b. The degeneracy gets lifted and you find two separated resonances. The T_D/T_B ratio has maxima of only 1/4. The width becomes much larger, and actually when you put the red frequency exactly on resonance, you get zero, i.e., no dark periods at all! Distinguishing between the weak and the strong intensity limits, you can see that this is really a quantum-mechanical system. And there are quantum jumps. But they're not quantum jumps between the states of a poor atom. They're quantum jumps between the states of the system atom plus laser fields.

Now here I'll tell you one more thing. In the original primitive figure (Fig. 1a), I showed you something that had convinced you all – and it convinced me completely too – namely the following: every one of the bright periods starts with a *red* photon, because now the electron has jumped down from the shelf. Because of quantum mechanics, that's not the way it always goes. That first photon can be *blue* sometimes too. The reason is that you can go from the red level to ground state by stimulated emission (i.e., with a photon that cannot be told from those of the laser). Thereafter you can absorb blue photons and go back and forth between G and E_b, and thus fluoresce in the blue; all this included in the formalism. This apparently strange behavior is just showing that we deal with a perfectly reasonable quantum mechanical system.

Several groups have done the experiment. Very few have done it with the simple three-level system I discussed here because the only such systems actually occurring in Nature happen to have their wavelengths in the ultraviolet and extraordinarily few people in this world have ultraviolet lasers – a small technical point. The people in Boulder, Wineland and company[3] do, and they were the first to do it in that manner. Toschek and collaborators[4] did it in Germany, but with more complicated (four-level) systems, and so have Dehmelt and his collaborators[5]. Toschek, et al., could show that you have fewer interruptions when you actually put *two* ions into the cage. Beautifully exhibited and qualitatively studied.

So there's only one more point left. And that is, to discuss carefully how these dark periods behave. And there one has to be very careful – not that it will have any practical consequences. The reason for that is that there are quite a few papers speaking about departures from exponential decay, and in fact there is a phenomenon called the "Quantum Zeno's Paradox" or "watchdog effect" [6]. Allegedly, when you keep observing a system all the time, you're bugging it to such a point that it no longer decays at all. It's a toy problem. There is even a paper that claims that proton decay has not been seen because we've observed it too long! Anyhow, if you want to go more deeply into the philosophy of these *dark periods as facts of measurement*, and I certainly think that one should, I could refer you to two moderately sane papers, both written by Italian people. (I think some of them are on the payroll of Abdus Salam, who is here and can certify that they are sane.)

This brings me to my conclusion, which I've already mentioned. I think that what will characterize the next twenty or thirty years will not be some enormous jump forward in particle physics, because the energies we are going to get are not sufficiently higher than the ones we already have today or shall have in the next year or two. Although there might, be of course, hordes of theorists working with the weird mathematics that we heard Professor Singer talk about, and of which he said all young people should know, (at which point I was for once extremely happy not to be a young person), there will be some anti-Copenhagen sanity recipes forthcoming, courtesy of Gell-Mann and Hartle, and also J. S. Bell.

Now, since you've been so patient, I'll show you one more of the more remarkable activities of Murray (the slide shows Murray, whip in hand, controlling a pride of tigers). Of course you must know that he never beats these animals; he's the only trainer who can use gentle methods.

Many happy returns, dear old friend!

References

[1.] H. Dehmelt, Bull. Am. Phys. Soc., **20** (1975), and also in *Advances in Laser Spectroscopy*, **95**, of the NATO Advanced Study Institute Series (Plenum, New York, 1983).

[2.] C. Cohen–Tannoudji and J. Dalibard, Europhysics Lett. **1** (9), (1986) 441. A first treatment, preceding the said Copenhagen conference but restricted to incoherent light sources, had been given by R.J. Cook and H.J. Kimble, Phys. Rev. Lett., **54** (1985) 1023. These authors, plus Ann L. Wels, presented a treatment of the coherent case in Phys. Rev. A, **34** (1986) 3190. The work of Cook, et al., eluded me when I was giving this talk.

[3.] J.C. Bergquist, R.B. Hulet, W.M. Haus, and D.J. Wineland, Phys. Rev. Lett., **57** (1986) 1699.

[4.] Th. Sauter, W. Neuhausen, R. Blatt and P.E. Toschek, Phys. Rev. Lett., **57** (1986) 1696.

[5.] W. Nagourney, J. Sandberg, and H. Dehmelt, Phys. Rev. Lett., **56** (1986) 2797.

[6.] An interesting discussion of this is given by R.H. Dicke, Found. of Physics, **19** (1989) 385.

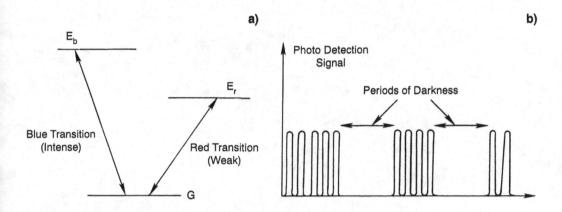

Fig.1 : (a) Assumed level scheme. (b) Sequences of (blue) fluorescence pulses; the first one in each train is generally red. Dark intervals correspond to "shelving" the electron in level E_r.

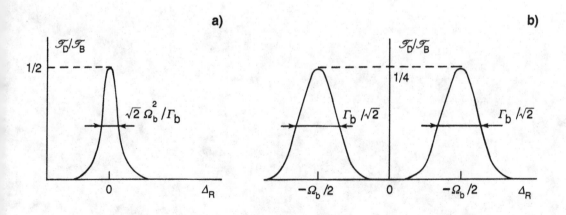

Fig.2 : The ratio of the mean lengths of dark and bright intervals vs. Δ_r, the detuning of the red laser. (a) Weak-intensity, (b) high-intensity limit.

The Gell-Mann Age of Particle Physics
Abdus Salam
International Centre for Theoretical Physics
34100 Trieste
Italy

I find I am three and a half years older than Gell-Mann although I have always prided myself on belonging to the same generation as he does. I shall give you a contemporary's views and some *early* recollections of Gell-Mann and his influence on the subject of Particle Physics.

I believe I first saw Gell-Mann at the Institute for Advanced Studies in Princeton in April 1951. He had brought from MIT the expression in terms of Heisenberg fields which would give the equation of the Bethe-Salpeter amplitude. I remember him and Francis Low working on this problem and producing the most elegant of papers, which has been the definitive contribution to this subject ever since.

I left the Institute for Advanced Studies in June 1951 and went back to Lahore. Later, in 1954, I returned to Cambridge and found that in the intervening period, the subject of new particles, the so called V^0-particles (Λ^0, K^0) had developed into a full-fledged new activity. There was the Gell-Mann-Nishijima formula which gave the connection between the charge, the isotopic spin and the strangeness - the prototype formula for other similar equations which followed this in later years and whose influence in Particle Physics one cannot exaggerate.

In July 1954, there was a conference in Scotland where Blackett took the chair and where young Gell-Mann was an invited speaker. After Murray's talk, I remember Blackett came up very fluid and said that this work of Gell-Mann justified his long hours of sleeplessness at Jungfrauhoch and Pic du Midi and the cold and misery of cosmic ray detection. I have never seen the granite-faced Blackett so emotional, particularly where theory was concerned.

Gell-Mann gave us a private seminar on what he then called deprecatingly "Mathematics" rather than "Physics". This was on the Gell-Mann-Low equations of the renormalisation group. The tremendous impact of this paper was felt much later. One may read about this in the beautiful account by Steven Weinberg in the Francis Low 60th birthday volume, where on page 1 he says "this paper is one of the most important ever published in quantum field theory".

At the 1955 conference at Rochester, I remember travelling back with Gell-Mann to New York. We sat at the Rochester Railway Station drinking coffee. I felt that the Σ^0 particle whose discovery we had heard about that afternoon from Steinberger should perhaps be identified with the Λ^0. Gell-Mann said to me that this could not be so because electrodynamics, which should provide the mass difference of masses between Σ^+, Σ^-, and Σ^0, would not account for the 70 MeV difference between Λ^0 and Σ^+ (because of factors of α). This simple observation was a good lesson for me not to economise too much with particles, in view of the simple consideration which Gell-Mann had pointed out to me - the first lesson I have (imperfectly) learned from Murray.

There was then the 1956 Conference at Rochester where, after the Goldberger talk, Gell-Mann came up with the suggestion of dispersion relations taking over in their entirety the subject of Particle Physics[1] - a suggestion of what later came to be known as the S-matrix theory by Geoffrey Chew. Earlier he, Goldberger and others had built up crossing symmetry and other desirable properties of four-dimensional field theory as a criterion which any decent field theory should possess. They used Heisenberg operators; that use was in contrast to what Schwinger and Dyson had taught us, which was to look upon the so-called interaction representation as the most desirable representation for field theories. I believe no one remembers today what the interaction representation was. This again I attribute to the influence of Gell-Mann and his co-workers.

Then came the great Gell-Mann period:

(1) The 1957 seminal paper on (V-A) theory with Feynman.
(2) An interlude with Regge pole theory followed by a succession of papers on symmetries. This culminated in:
(3) The 1961 work which generalised the isotopic SU(2) to SU(3) flavour symmetry.

---

[1] I was interested because with Walter Gilbert - my good pupil at Cambridge who went on to win the Nobel Prize for identifying residues on the double helix chain in 1980 - we had also written down the dispersion relations which formed the basis of this work with Goldberger. The famous Gell-Mann quote of the pheasant being cooked between the pieces of veal (while the veal is discarded at the end) was probably meant already for this suggestion of his. This is the second lesson which the world has learned from Gell-Mann.

(4) The prediction of the Ω^- in 1962. This was parallel to the work done at Imperial College on SU(3) and Ne'eman's work on the 8-fold way, with the difference that Gell-Mann understood clearly that if the isotopic spin group was considered as SU(2) rather than as SO(3) (as Nicholas Kemmer, the originator of the isotopic group at Imperial College had taught us), SU(3) would follow on SU(2).

(5) In 1961 the famous Gell-Mann current-algebra relations were stated at the conference in La Jolla.

(6) In early 1963 came the paper which described the "mathematical quarks" as Gell-Mann had always called them - signifying their permanent confinement.

In 1967, at the conference in Heidelberg, Gell-Mann gave the final summary talk. This was devoted to showing that the saturation of left-right symmetric SU(3) x SU(3) current-algebra relations did *not* provide the desirable physical particle spectrum. The audience loved this demonstration on the part of Murray that he could be fallible - though still a master.

I think it was 1967 that Marshak held a meeting in Rochester at which Feynman paid the ultimate tribute to Gell-Mann, The topic was "What ideas I have found most valuable in Particle Physics" and Feynman's reasoned and memorable intervention was to say that it was "Gell-Mann's work". Marshak has told me that Gell-Mann was not there. This, I think, was the ultimate accolade which Gell-Mann could receive from a contemporary of the standing of Feynman.

In 1969 came the highly deserved Nobel Prize, which might have been given to him any time after 1955.

So far as my own work is concerned, Gell-Mann has always been very forthcoming. I may mention one occasion as an example. This was when I mentioned the possibility of proton decay in December 1973 - which Pati and I had earlier suggested - at the Conference at the University of California at Irvine, organised by Reines. Gell-Mann had not heard of this idea before. He accepted it immediately and remarked on it publicly the next day in his talk.

Since 1969 Gell-Mann has been spending his time not only in physics (where he has made important contributions to the ideas of asymptotic freedom, to elucidation of QCD and to grand unification as well as to string theory) but has also found time for global concerns.

Gell-Mann has defined an "Age of Particle Physics". May he live long to give us guidance further, not only in Particle Physics, but also in other human concerns - as a lasting testimony to his uncompromising search for truth and excellence.

Remarks on the occasion of Murray Gell-Mann's more or less 60th Birthday

M. Goldberger
Institute for Advanced Study
Princeton, NJ 08540
USA

I first met Murray when he was a small child, a 19 year old graduate student at the Caltech of the East. I was an ancient of 26 at the time and was quite surprised when he announced upon meeting me that he knew who I was and that he had read all my papers. That was not such a monumental task at that time, but I found out that he had indeed read them. I discovered much more quickly than Viki Weisskopf that Murray was different from me and thee. We became friends and have remained so for nearly forty years.

When I went to Chicago in 1950 I began immediately agitating to hire Murray. It was no easy task to convince my senior colleagues that this was sensible since he had identically zero publications to his name. I did, however, prevail and we began a long collaboration that continued episodically for nearly 20 years. This was an exciting time in particle physics when there was a vast amount of experimental data and a paucity of theoretical tools to cope with it. It was a pleasure to work with Murray as we used everything we could lay our hands on theoretically to try to pick our way toward an understanding of what was a bewildering and complex landscape. His ingenuity, intensity, enthusiasm, and confidence that we could understand a great deal if we stuck to general principles and were not afraid to make bold conjectures was contagious.

Murray's domination of the entire field of elementary particle physics was awesome. In 1966 when we were planning the high-energy physics conference at Berkeley, we thought to have a series of introductory talks on theory, strong interactions, weak interactions, etc. and then began to argue about who should talk about what. After much discussion I offered a trivial solution: Let Murray do it all! He did and it was brilliant.

Those of you whose memories are relatively intact probably have noticed that Murray's is incredible. I can't tell you how many times I have been harangued by verbatim repetitions of responses to questions or to stimuli like ... I almost hesitate to mention some of them for fear that he will uncontrollably break into one of his perorations - archaeometry, failure of traditional policy studies, the Santa Fe Institute, just to mention a few. And, of course, his recall of obscure and abstruse facts going back to literature like *The Little Engine that Could* or even earlier is legendary.

I'd like to make a few remarks about Murray's relationship with Caltech. In late 1954 he left Chicago to visit Columbia, as I recall, and after some flirtation with them made a trip to Pasadena; shortly after a few days' visit during which time he dazzled the natives, he was made an offer and he has been at Caltech more or less continuously ever since. Except for a brief honeymoon period he constantly found much at fault with the institution. But even when Berkeley, MIT, The Institute for Advanced Study, and many others tried to lure him away, he somehow always said no. Finally Harvard called and Murray said yes. I remember that it was around Christmas time, I'll say 1961 but I'm not at all sure of the year, Murray called to tell Mildred, my wife, that he was going to Harvard. The more he talked about it, the more apparent it became to us that he was really sorry that he had accepted the offer. Finally Mildred said "Murray, you don't have to go; they can kill you, but they can't eat you!". And so it came to pass that he sent the Harvards a telegram. Frankly I think he made the right decision. I think it is time to resign yourself to the fact that you will always be at Caltech, Murray. I can think of worse fates.